WHOSE LAND IS IT ANYWAY?

A controversial study of the health of the British Countryside, and an assessment of present policy towards agriculture, nutrition and la~~~ ~~~

M Kennedy
3 Buckingham St
Scunthorpe

WHOSE LAND IS IT ANYWAY?

Agriculture, Planning and Land Use in the British Countryside

by

RICHARD NORTON-TAYLOR

diagrams by Jan Hoult

TURNSTONE PRESS LIMITED
Wellingborough, Northamptonshire

First published 1982

British Library Cataloguing in Publication Data

Norton-Taylor, Richard
    Whose land is it anyway?
    1. Land use, Rural—Planning—Great Britain
    I. Title
    333.76'0941   DA667

    ISBN 0-85500-095-3

Printed and bound in Great Britain

CONTENTS

for Hugo and Sam,
and their generation

LIST OF FIGURES

LIST OF PHOTOGRAPHS

Acknowledgements.

Throughout the time I was writing this book, I was encouraged by the growing, but in my view still insufficient, debate about what is happening to our land and the increasing threats to it. I would like to thank colleagues, officials and farmers who have helped me and provoked me whether or not they realised it at the time. In particular, I would like to thank Christopher Robbins, formerly of the Centre for Agricultural Strategy at Reading and now director of the Coronary Prevention Group, who provided me with material and ideas and who made it clear to me that agricultural policy in Britain is made – or simply allowed to develop – without any consideration of other issues, including diet, health and land use. The work of Alice Coleman, of King's College, London, was also invaluable. Thanks are also due to Jan Hoult for the maps and diagrams.

But this book would not have been written without the vision and enthusiasm of Alick Bartholomew of Turnstone Press. He was an essential source of guidance, inspiration, and encouragement. And I would not have survived without the understanding and patience of my wife, Anna.

Permission to quote from John Steinbeck's *The Grapes of Wrath* was given by William Heinemann Ltd. For providing photographs: to *The Guardian* for those on pp. 62, 84, 93, 112, 173, 208, 213, 225, 234, 255, 260, 264, and 280. Credits are otherwise given next the photographs.

INTRODUCTION

"But how can land be cultivated when there is nobody to cultivate it? 'We have fields; men go by, but never go in,' an old labourer said to me. And so it is in reality... The British nation does not work on her soil."
  Peter Kropotkin, *Fields, Factories and Workshops.*

The fields flash by as we look out of the train window. We catch a quick glance at the countryside as we drive hurriedly to escape from city life and work. What's going on out there?

Land is our only permanent resource. Its produce, if nature was allowed to have her way, is infinitely renewable. Yet it is abused, wasted, treated just like any other commodity. It is at the mercy of individual wealth, greed, mere complacency. It is prey to attitudes that put short-term profits before the longer-term interests of the community. Thus farmland is drenched with chemicals to produce higher-yielding crops; farmers dispense with labour to channel their earnings, loans and grants into pesticides and machines. Farms get bigger and bigger so there are fewer and fewer. Wealthy individuals, companies and financial institutions move in as smallholders cannot keep up with the race, just as corner grocers give way to supermarkets filling up more and more shelf space with processed and convenience foods.

The land is a source of glaring anomalies that lie at the heart of many of Britain's problems: the crisis of housing, of a callous bureaucracy, of planning, of pollution. Industrial companies, claiming poverty, pollute the environment, and Whitehall turns a blind eye to legislation designed to protect our health, as well as that of the land and of wildlife. There are thousands of empty houses, yet the waiting list for homes grows every year. In three years in the early 1960s, 110 tower blocks were built in Liverpool; many of them are now modern slums in a deprived environment, victim of a cold, paper dream shared by bureaucrats, planners and politicians. In rural areas, opulence and deprivation stare each other in the face.

Land is abandoned and ignored. Though people are now moving out of the large cities, especially London, to more open spaces, such as East Anglia and the South West, about half the population of Britain still lives within a 100-mile radius of the capital. Rigid and inconsistent attitudes and planning laws have turned villages into mausoleums or a mixture of tourist traps and second homes. More and more of us seek refuge in the country, and more and more people read about the country, nostalgically perhaps, in country magazines which are enjoying their most successful period ever. Government policies, towards planning, public services, even food, have been dominated by an urban bias, though the state of the inner cities scarcely suggest it. Meanwhile, the fight is on to save one of the most beautiful man-moulded landscapes in the world.

The best farmland has been developed for urban or industrial uses at a rate five times as fast as poor land, yet there are half a million acres of land lying idle in inner cities[1]. New roads eat up precious farmland, while food — the erstwhile product of that land — accounts for more road traffic than any other commodity. While millions were being squeezed into cities, the Northumberland National Park lost two-thirds of its farms and farmworkers in twenty years, and the Lake District lost a quarter of its agricultural workforce and a fifth of its farms. While unemployment increases relentlessly, more than twenty-five percent of the country's entire railway network will have to be replaced by 1990 or become unusable. Our canals and many waterworks remain clogged up.

Farmland, stretches of river, woods, moors, islands, abbeys and castles are offered and sold on the market as though they are ordinary consumer durables. In September 1980, one estate agent announced that a 3,000-acre estate near Edinburgh was on the market. "One of the many attractions (of the estate)," it said, "is that it is run with only 2 full-time employees, a shepherd and a keeper." A year later, the agents enthusiastically proclaimed that over a period of three months, it had organised sales of more than 12,000 acres for a sum close to £12 millions.

In 1982, a private offer had been accepted by the existing owner, John Stancer, a Leamington Spa solicitor, for 6,000 acres of Farndale in the heart of the North Yorkshire Moors National Park. The Park Committee did not know who the prospective buyer was. But it did know that he was prepared to pay about £1.5 millions for the estate which included 22 tenanted farms, a house with nine cottages, 2 complete hamlets, a 2,000-acre grouse moor, a pheasant shoot, a trout stream, and nearly 200 acres of forest.

David Goldstone, a Welsh solicitor, had just bought Land's End, tiny compared with Farndale, but important symbolically — for £1.75 millions. At about the same time, Victor Watkins, another Welsh millionaire and owner of a number of London property companies, bought the Cotswold village of Salperton, near Cheltenham, for about £3 millions. The village includes 30 cottages, a 1,600-acre estate, and a pheasant shoot. The village was sold by Sir Edward Hulton, who paid £50,000 for it in 1950.

The North Yorkshire Parks Committee failed in its attempt to persuade the Government, through the National Heritage Fund, to bring Farndale into some kind of public ownership. The National Trust could not afford the money to buy Land's End.

When it pushed its Wildlife and Countryside Act through Parliament in 1981, the Thatcher Government ensured that in future landowners would be compensated if they agreed not to develop, drain, or plough up habitats or stretches of traditional countryside but, instead, preserve them for the community as a whole. This is a dangerous precedent and an arrogant principle; it is also unworkable since conservation agencies, notably the Nature Conservancy Council, have neither the financial resources, nor the confidence, to pay the farmers who continue to benefit from generous grants from the Ministry of Agriculture to develop their land. The battle is already lost; the first test cases involving the Halvergate Marshes in Norfolk and the Romney Marsh in Kent proved the point: early in 1982, the conservationists were defeated, the landowners won. Some farmers are even now putting to the plough land that they suspect might be put under conservation order, lest they lose their freedom to do what they want with 'their' land.

The Act ignored the persistent claims by farmers that they own the land in trust for the whole nation and for future generations, with the clear implication that they have no right to use the countryside merely as a source of maximum, short-term profit.

Agricultural landowners have fought successfully against any proper investigation into landownership. They resent interference from Whitehall though they demand financial guarantees, subsidies and grants from the Government, the EEC and taxpayers. "The farmer, like the businessman, hated filling in forms... He grumbled, as he had always done, but he knew that state intervention had brought him real prosperity."[2] The commentator was describing the dramatic changes that took place in British agriculture during the Second World War, but the sentiment remains true today. The Thatcher administration bowed to it by telling farmers that, in future, they can give even less information about their crops and what they produce and that they can go ahead and plough up land without waiting for grants or advice about whether their plans are in the best interests of the countryside.

The land is under threat from the relentless search for new energy sources, including coal and oil. It is under pressure from an agricultural system which is increasingly dependent on expensive capital equipment, energy and chemicals. This, in turn, has meant that traditional landowners who do care for the countryside are having to give way to a new breed of business-farmers whose main interest is to produce as much as possible as fast as possible.

It is remarkable, given the vital importance of land and food, that successive British Governments have never drawn up a coherent agricultural and land policy. Little or no thought is given to the conflict over the different types of landownership, what kind of crops should be encouraged to promote greater national self-reliance, the relationship between agriculture and nutrition. Peter Walker, the Agriculture Minister in the Thatcher administration paid lip-service to the principle of self-sufficiency, yet farmers are encouraged to produce food for which there is no market at home, simply because of the Common Market's open-ended commitment to pay farmers high prices for whatever they produce. Britain's food exports are rising to record levels, stimulating an international agricultural trading system that is increasing the dependence of poor countries on the rich. According to World Bank estimates, there are 800 million people living in absolute poverty, yet we in the West are producing mountains of surplus food at prices the poor cannot afford. We are imposing a food system based on our own expensive, irrational criteria and double standards. The US, exports pesticides and other chemicals to the Third World though they are declared unsafe for use at home; the Amazon jungle is sprayed with defoliants bought by Brazil from the US at cut price after the Vietnam War. The objective: to set up cattle ranches to produce meat for the rich.

Back home, milk production increases even though the Government itself suggests that we should cut down on our consumption of fat. The Government negotiates an EEC subsidy on butter so that it can claim consumers do get some financial benefit from the Common Market — not because it is good for you. Butter prices still increases, so consumption continues to fall. The Government negotiates an EEC subsidy for school milk but ignores the views of its own advisers by refusing to allow the subsidy to be used for skimmed milk. The Government says it believes in free choice, but uses tax policies and other instruments at its disposal to promote the interests of large food companies.

We are seduced by the superficial attractions of conspicuous consumption and of a wasteful, throw-away, society. We are seduced by experts, scientists and technicians who cannot see the wood from the trees. We bow to unsophisticated, material-istic indicators and measurements such as the Gross National Product or notions of 'cost-effectiveness'. How can and how should Britain, with its ailing economy, survive in a highly competitive world which still worships the concept of economic growth? What are the social and environmental costs, or the costs in terms of the potential hazards involved, of the nuclear power programme?

A few days after agreeing to invest in a new and costly torpedo for the Navy, the Government in 1981 refused to guarantee the financing of a gas-gathering pipeline that would have harnessed an energy resource that literally goes up in smoke every day over the North Sea.

North Sea oil may give the impression that Britain is secure to maintain its traditions as an important trading nation, traditions which were perpetuated by the conquest of other peoples' land and the resources — human, agricultural, mineral — it produced. But the reality is that we are now becoming relatively poor and vulnerable in a world that does not owe us any favours.

Only by a change in values and priorities can we restore some sort of balance to the way we treat our environment as well as redress inequalities and redistribute land and property — inequalities that are likely to become a growing source of tension and unrest as more of us search desperately for space.

Society is becoming dangerously polarised. Some of us may choose to reject conventional values, only to opt out to a small place in the country and try, ostrich-like, to become self-sufficient, forgetting what is happening over the hill. We should not turn our backs on technology, but we must use it, not allow it to abuse us. There is an alternative to accepting an increasingly bureaucratic and secretive technocracy, with Governments spending thousands of millions of pounds on unemployment benefits, and paying 60 times more equipping soldiers than educating children.

Modern farming methods are becoming increasingly irrational and unnecessary. They are destroying the countryside, as well as such visible links with our past as ancient sites, hill forts and earthworks. Without a change in priorities and values, unless we question our devotion to material consumption, we shall not only be out of tune with our environment, but also ill-equipped to face the rapidly-changing economic and social conditions that are enveloping us.

1

THE OWNERSHIP OF LAND

"Land always carries associations – of status, security, rights – more profound than the value of the crop."

E.P. Thompson, in The Making of the English Working Class.

Defence of 'private property' is deeply engrained in British law and culture. In the British philosophical tradition, reflected best perhaps by John Locke, the possession of property is treated as a fundamental, natural right. "The Englishman's home is his castle" is a time-worn cliché but no less appropriate for that. The concept of 'freedom' is closely allied to property ownership and the notion of a property – owning democracy. And in Britain there are no statutory restrictions on who can buy land, or the size of an estate or farm. In practice, it goes to the highest bidder, or else it is the gift of inheritance. For the easiest way still to acquire land is to be born in the right bed.

Landownership has been a source of power throughout British history. At first, it was the most important instrument of patronage in the hands of the monarch and was extravagantly used by William the Conqueror and Elizabeth I. It was also brutally used by Oliver Cromwell, though by the time of the Civil War half the agricultural land of England was already enclosed, that is identifiable as 'private'. The seventeenth century saw the rise of the merchant classes. By the eighteenth century, agricultural development was inspired less by an altruistic desire to feed a growing population than by the desire to fatten rent rolls and large profits[1]. By the nineteenth century landownership secured its place as a status symbol and source of wealth, and the traditional landed aristocracy had to begin to share it with industrialists.

Relatively recent members of the peerage, with titles created between 1916 and 1922, include: Viscount Cowdray (one of the richest men in Britain and head of a financial empire that includes a merchant bank, Lazard Bros., the publishing group, Pearson-Longman, and *The Financial Times*); The Earl of Iveagh (Guinness) with about 24,250 acres in Norfolk, and Viscount Leverhulme (soap) with an estimated 100,000 acres in Cheshire.[2]

The Vestey family, also among the very richest in Britain, is a testimony to how industrialists and businessmen became both peers and large landowners. The family was rewarded with a title in 1922, having supplied British troops with cheap beef. Its recent claim to fame arose out of the exposure by Phillip Knightley in *The Sunday Times* in 1980 into how the Vesteys carefully used loopholes in tax laws which enable overseas trusts to escape the clutches of the Inland Revenue. In this way, the Vestey family, with trusts based in Uruguay, have been able to avoid paying millions of pounds' worth of income tax. The family presides over one of the most private and largest businesses, with interests ranging from insurance companies to guard dogs, and including the Blue Star and Holt shipping lines. Its Union International Company alone has assets conservatively estimated at £50 millions. It controls 20,000 square miles leased from the Australian Government, owns 2,000 square miles in Brazil and another 1,000 square miles in Venezuela. It has a large stake in the New Zealand meat trade and owns the Dewhurst chain of butchers.

The family also persuaded the Government to exempt it from Capital Transfer Tax on most of its 100,000 acre estate at Assynt in Sutherland, in the North of Scotland, in return for agreeing not to develop it and to allow reasonable access to the public. Most of the estate is used for shooting. Lord 'Sam' Vestey is an enthusiast for polo, like Viscount Cowdray, and a friend of the Royal Family. The Vesteys are perhaps an archetype of the family which made its money in commerce and then took on many of the attitudes and landed assets of the traditional aristocracy.

For many business leaders farmland has been seductive, though sometimes dangerous, investment. One spectacular deal which upset local farmers was made in October 1972 when the Brown Candover estate in Hampshire, covering 2,350 acres, was sold to Sir John Reiss, the chairman of Portland Cement, for £3.2 millions. A month later, Reiss sold the estate to Ronald Lyon, a wealthy property developer, for £1,400 an acre — an exceptionally high price at the time — leaving Reiss with a profit of more than £100,000. In the course of the year, Lyon bought a total of 10,500 acres at a cost of £7 millions. But the property boom slumped, Lyon discovered that farmland could provide only a long-term capital investment, his group collapsed, and he sold off his farms to the Post Office Superannuation Fund.

But politics and landownership have been more faithful bedfellows, at both the local level and at Westminster. Sir Horace Cutler, former Conservative leader of the Greater London Council, is joint owner of a property development company and of a farm near Farnborough, Hampshire. After a deal in which he sold to the local council part of the farmland at about three times the price which he had originally paid for it, Cutler acknowledged in an interview with the London magazine, *Time Out*, that he was worth on paper between £1 million and £2 million and was not embarrassed by the sale. "It's not my price;" he said, "you have to understand that. It's been set by the District Valuer . . . my companies are not philanthropic organisations."

Mrs Thatcher's Cabinets, the wealthiest for a very long time, included many traditional landowners:

Lord Carrington, Mrs Thatcher's first foreign secretary, is one of the richest members of the Tory landed gentry, with about 1,200 acres in Buckinghamshire as well as land in Lincolnshire.

Willie Whitelaw, Home Secretary, was left by his grandfather estates in Dunbartonshire on rich coal seams. He is a partner in a farming company in Penrith, his constituency. In the House of Commons Register of Members' Interests, he adds: "I still retain a comparatively small amount of land near Glasgow, the ownership being mainly in the hands of my daughters".

Francis Pym, owns about 500 acres in Bedfordshire, inherited from his family, including the estate's Georgian mansion, Hazell's Hall. Michael Heseltine, the Environment

Secretary, farms 400 acres near Banbury.

James Prior, owns 380 acres of farmland near his constituency of Lowestoft. John Nott, owns a 130-acre farm in Cornwall, and Sir Ian Gilmour, who had been Lord Privy Seal until 1981, owns land in Middlesex. He married into the Buccleuch family, Britain's biggest private landowner. Michael Jopling, the Chief Whip, owns a farm in Westmoreland. Adam Butler, a Minister of State, is a Warwickshire farmer. His brother, Richard, is president of the National Farmers' Union. They are the sons of 'Rab' Butler, the eminent Tory politician who died in March 1982. Peter Walker, Agriculture Minister, farms 200 acres in Shropshire. Two of his junior Ministers are also farmers: Alick Buchanan-Smith inherited land in Midlothian, Earl Ferrers, land in Norfolk.

George Younger, the Scottish Secretary, has a share in the family's 1,000 acre farm in Stirlingshire, and is son and heir of Viscount Younger of Leckie, director of Tennant Caledonian Breweries. Lord Mansfield, Minister of State at the Scottish Office responsible for agriculture, inherited a 33,000-acre estate in Perthshire. Sir Hector Monro, at one time Minister for Sport, owns a farm in Scotland, and Lord Thorneycroft, former Minister and Chairman of the Conservative Party until 1981, owns 44,000 acres in Scotland.

The former Prime Minister, Jim Callaghan, is part-owner of a 200-acre farm near Lewes in Sussex, though he declares that he derives no income from it. He also owns an attractive farmhouse next to the farm. His property was valued in 1978 at £120,000.

Many leading politicians — Denis Healey and Roy Jenkins among them — own houses in the country.

In the EEC context too, landowners are much in evidence. The Tory group at the European Parliament includes at least sixteen of them, more than a quarter of the total. It is appropriate, perhaps, given that the Common Market budget gives more to agriculture than to any other sector of the economy. They include the Marquess of Douro, son and heir of the Duke of Wellington and Euro-MP for Wessex, and Mr James Scott-Hopkins, who comfortably captured the Hereford and Worcester seat. Sir Fred Warner, a former Foreign Office

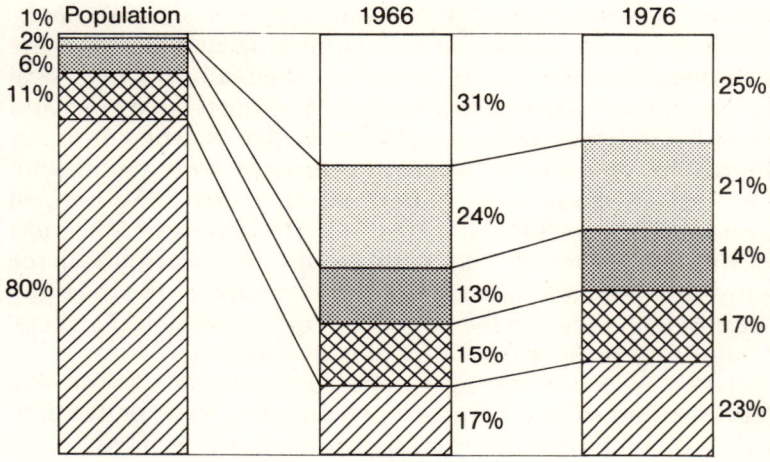

1 a. Distribution of wealth: *the first column represents the total population, the second and third all individually-held wealth. Since the bulk of private property and assets are distributed among the immediate family before the death of the owner in order to avoid taxes, the rich tend to remain wealthy.*

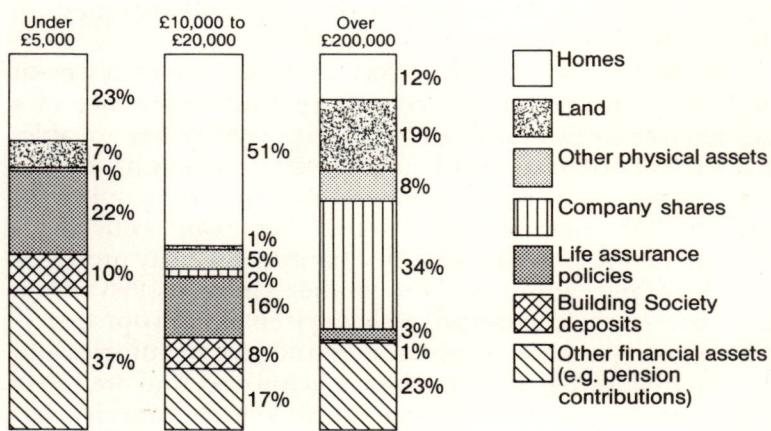

1 b. *Those with a high annual income own the most land. (Source: Roy. Comm. Distribution of Income & Wealth).*

diplomat, is the Euro-MP for Somerset and a farmer. Miss
Beata Brookes, a landowner, won North Wales for the
Conservatives after a campaign in which she warned local
farmers of the danger of Europe's communists taking the land
away from them. Sir Brandon Rhys-Williams, who combines
his seat in Westminster, where he represents Kensington, with
his seat in Strasbourg, where he represents South-East
London, owns a large slice of the Vale of Glamorgan. The most
prominent farmer in the Tory group and chairman of the
European Parliament's agriculture committee, is Sir Henry
Plumb, who resigned the presidency of the National Farmers'
Union to represent the Cotswolds in Europe.

There is an interesting conundrum whereby if you own
land, you presume — and are presumed — to be better fit to
rule, to hold power and responsibility, while if you have
successfully climbed the political and industrial ladder, then
you should own land. And those who invest in farmland, like
the Victorian industrialists before them, (helped by the
agricultural depression and low cost of land), often attract the
support of others with similar ambitions. Meanwhile the
aristocracy and landed gentry enjoy that added appeal,
peculiar to Britain, of a paternalistic Whiggism, a kind of
enlightened, though frequently arrogant, belief that they are
the backbone of Britain.

The distribution of income has been the subject of a great
deal of controversy over recent years. But the absence of a
debate over the ownership of land and property is remarkable,
given the concentration of landed wealth in so few hands and
that it is such a visible source of wealth, including unearned
wealth. "Throughout the period we can collectively
remember," noted A. H. Halsey, Oxford University professor
of social and administrative studies, in his 1978 Reith
Lectures, "three-quarters of us have been virtually property-
less in that area which covers the central part of our lives and
our occupations — how we earn a living and how we relate
ourselves to our fellow men . . . A tiny minority has
monopolised wealth, and an even tinier minority has
monopolised property for power."

He was describing all kinds of property but the inequalities
of landownership are even more striking. Just over 1 percent

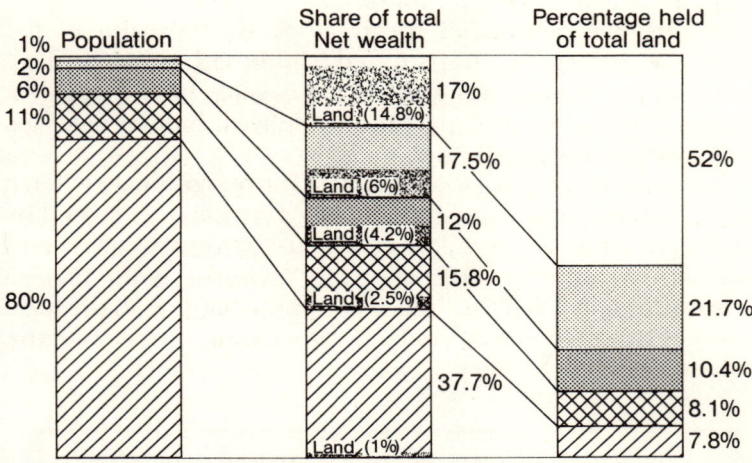

2. **Land Ownership:** *the figures referring to land in the middle column relate to the proportion of total wealth of that group held in land. The Country Landowners' Association has won many tax concessions for farmland owners. (Source: Roy. Comm. On Distribution Income & Wealth)*

of the adult population owns almost seventy percent of the land[3]. The concentration of landownership in Britain, encouraged by the Enclosures and the Industrial Revolution, is unique in Europe. Elsewhere, political and social revolutions led to a more drastic break-up of large estates. On the Continent, the aristocracy and traditional landed classes failed to hold on to their interests; in Britain, they have a subtler instinct, they knew how to compromise politically. And in Britain the aristocracy was shielded from the Napoleonic Code whereby property is divided among the family; primogeniture continued in Britain where property was handed down to the eldest son.

Inheritance therefore played a much more significant part in the pattern of landownership in Britain than in other countries. It still does so: the relative decline in ownership of wealth enjoyed by the top 1 percent has been picked up by the group (the 2-5 percent of the most wealthy) immediately below the very top.[4] Because the bulk of the property is distributed among the family before the death of the owner often so as to avoid taxes, the wealth of rich families has been distributed among rich families.

Seventy-five percent of the total personal wealth of the richest 1 percent is transmitted, inherited, wealth. This accounts for more than half of the wealth of the top five percent. The bottom 50 percent in the wealth table own less than one percent of all land.

The concentration of wealth is even more significant when it comes to farmlandownership. Britain's farmland is owned by just 2 percent of the population. The trend towards fewer and fewer farms continues: 10 percent of all farms — the biggest — produce half of all the food grown in Britain. Half the food grown in Britain comes from land owned by 0.2 percent of the population.

"Knowledge of ownership is a matter of hearsay, and local gossip, rather than any statutory and central register of landownership. A lovely anachronism of England. Long may it remain" – *Peter Wormell, farmer of 900 acres in Essex, former member of Essex County Council, the Water Board and his local rural district council.*

"The paucity of comprehensive up-to-date information on land ownership is remarkable" – *7th report of the Royal Commission on the Distribution of Income and Wealth.*

Landownership carries tremendous responsibilities. But it is also a privilege enjoyed by a few. And secrecy, as Whitehall has demonstrated so well, is one of the most effective protectors of privilege.

Britain is one of the few countries in the world which does not have an open and comprehensive system of land registration. The Ministry of Agriculture itself does not know exactly how much farmland there is, and therefore how much is lost to industry and urban development, or roads, nor who owns it, nor how it is managed. As David Rose of Essex University observes: "The National Farmers' Union and the Country Landowners' Association both complain, quite regularly, about what this or that Government policy means in terms of a threat to landownership. If they oppose systematic investigation of the subject, how can they expect Governments to pursue what they would regard as sensible policies in this area?"

Without this information Parliament and elected authorities can neither control the pattern of ownership, nor even monitor it. The limited information that is available from district valuers or the Land Registry is a monopoly allowing privileged access to information to a small number of officials. It is one of the factors which makes a mockery of the argument that public inquiries are fair. Objectors and the public do not have the resources available to Government while the Government can disclose what information, known only to itself, it wishes to disclose and what information to withhold.

No sensible debate about the land can be conducted without knowing who the landowners are. Yet landowners enjoy a great deal of discretion — and the Government, absolute discretion — over how they use the land, in spite of planning legislation. Adequate information about land-ownership is also important if elected bodies are going to be in a position to monitor the effects of taxation, tax allowances, subsidies, grants and guaranteed prices for farmers and developers.

In many European countries, land registration was inspired by that bureaucratic genius, Napoleon. Their range varies — in West Germany, the register goes as far as to include soil classification. In Britain, Scotland is in the vanguard. There the Register of Sasines is open to the public, although the statutes are difficult to interpret and no distinction is made between the legal, formal owner and the real beneficiary owner. Even for those who are allowed access, it is not possible to establish individual ownership by referring to the Land Registry of England and Wales. The process of detection is even more difficult when banks act as trustees. While they are the legal, registered owners, in practice the land may be owned by financial institutions such as insurance companies or pension funds. Off-shore landowning companies registered in the Channel Islands, for instance, or Gibraltar, can escape the provisions of the 1976 Companies Act whereby the beneficial ownership of shareholdings of more than 5 percent has to be declared.

As part of its attempt to soothe the Scottish Nationalists, the Labour Government in 1978 agreed to set up a more comprehensive and accessible land register for Scotland. Asked why a similar register would not be introduced for the rest of Britain, the Government explained that traditionally the issue of landownership was more controversial in Scotland than in England or Wales and that, in any case, it would be too costly to extend such a system throughout the country.

"Landownership", a Scottish Office Minister told the House of Commons in December 1977, "would be costly to public funds." And he added: "whether it would be a waste is a matter for debate." The Government, and Whitehall, was confident that the debate would not take place. The initiative to set up an inquiry in 1977, under Lord Northfield, into the acquisition and occupancy of agricultural land, was the result an off-the-cuff reaction by the then Minister of Agriculture, John Silkin. He was responding to the growing controversy over the specific issue of land purchase by financial institutions and foreigners. The initiative was not the result of a considered decision following serious appreciation of the problems of land use and ownership.

When the Ministry of Agriculture announced that on January 1, 1978, it was going to launch a pilot study of farmland ownership by looking at the Wyre Forest area on the borders of Hereford and Worcester county and Shropshire, it did so almost apologetically. The confidentiality of the questionnaire, it insisted, would be respected. Even so, over a third of the landowners refused to cooperate.

In 1978, Peter Shore, then Environment Secretary, rejected a suggestion that local authorities should be obliged to publish a register of all their landholdings. But a year earlier, the Civic Trust succeeded in conducting a survey of dormant land in England and Wales, and found 250,000 acres, after the Department of the Environment admitted that it had no reliable figures and insisted that a proper study could be undertaken only at 'disproportionate cost'. Cost is one of Whitehall's more familiar excuses — the argument with which it tries to seduce popular support for its passion for secrecy. More openness, it says, would be expensive.

When the Conservatives were returned to power in May

1979, Shore's successor, Michael Heseltine, was determined to free land from many planning controls and promote private development. Thousands of acres of dormant land, especially in cities, are owned by local authorities. To encourage them either to sell or develop, Heseltine decided that all dormant land held by them should be listed in a public register.

Heseltine had a political motive. Ironically, political motives, especially on the Left, are political most frequently cited by private landowners as one of the dangers of providing information about their agricultural property. It would be a prelude, they argue, to land nationalisation or a tax on wealth, even though this has not been the case in other countries. The objective benefits of having more information about landownership were spelt out by the Northfield committee: "The lack of data on agricultural landownership," it said, "is in sharp contrast to the wide range of statistics upon which agricultural support policy is based. Yet Government fiscal policy may well affect the industry more fundamentally than annual changes in (farm) prices." How can we tell whether taxes are having the desired or, indeed, any effect without more information on landownership? For straightforward agricultural reasons — to help adopt a coherent agricultural policy for Britain — the committee said more knowledge is needed "on almost every aspect of landownership, acquisition and occupancy if Government policy is to be formulated on an adequately informed basis".

The land is our birthright and the root, the origin, of the State. By the same token, the State is nothing without the people in it; it is meaningless without consideration and fulfillment of the needs, wishes and health of its inhabitants. The more ignorant we are about the land and how it is used and by whom, the more we are divorced from what is ours. So in turn, and that is sufficient reason for the plea for more information, Government secrecy promotes the concept of a State with a personality and power of its own.

The first official survey of landownership in Britain — the Domesday Book, completed in 1086 — was commissioned by William the Conqueror. It was inspired, latter-day critics say, by the bureaucratic mind of an occupying Norman. The second. and latest, survey was in 1874 Return of Owners of Land, commissioned by Lord Derby and intended to demonstrate to the reformists among the Opposition then that ownership had spread much wider than they alleged. To Derby's embarrassment, the 'new Domesday survey', as it was called, showed that about 75 percent of the country was owned by 7,000 families.

If we based our knowledge on information offered by landowners or by the Government, we would have little idea about who, and how many, owns the land today. Those who do know are not prepared to share their secrets. According to George Inge, a director of Savills, one of the leading estate agents: "there are ten land agents, ten individuals, in Britain who could tell you who owns every acre in the country." Ask

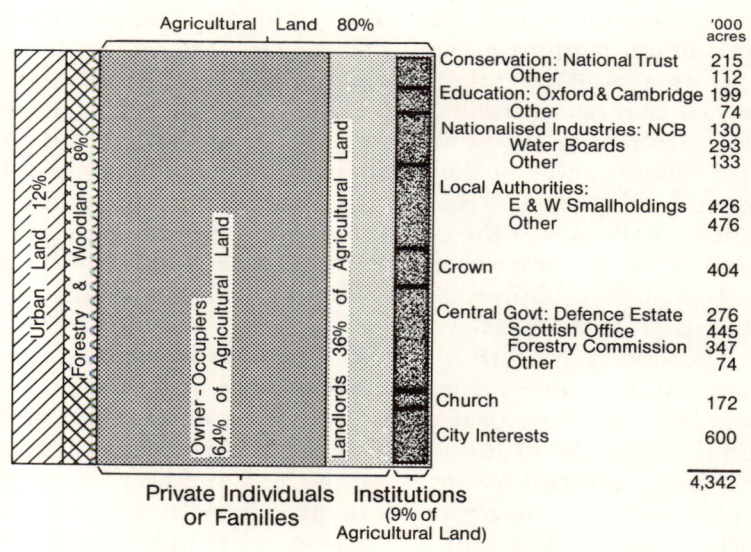

	'000 acres
Conservation: National Trust	215
Other	112
Education: Oxford & Cambridge	199
Other	74
Nationalised Industries: NCB	130
Water Boards	293
Other	133
Local Authorities:	
E & W Smallholdings	426
Other	476
Crown	404
Central Govt: Defence Estate	276
Scottish Office	445
Forestry Commission	347
Other	74
Church	172
City Interests	600
	4,342

3. **Ownership of Farmland:** *the maze of landownership is hard to unravel, partly because of the growth of trusts, partnerships and farming companies. About 1,300 individuals own about one third of Britain. But the large financial institutions are steadily increasing their stake in the best farmland. If they so wished, they could control about half Britain's total food production within 30 years. (Source: Northfield Report, 1979).*

him who these individuals are: "I'm afraid I cannot tell you that," he replies; "that is confidential."

We can, however, estimate that on the basis of available information and research:

Traditional institutions, such as the Church and the Crown and public and semi-public bodies, including Government departments and nationalised industries, own 4.27 million acres; or about 9 percent of the total agricultural area.

Financial institutions, including insurance companies and pension funds, own 530,000 acres, or just over 1 percent of all farmland and 2 percent of the area under crops and grass. By a process of elimination, private individuals, either directly or through companies and trusts, own about 39 million acres, or 90 percent of all agricultural land, of which foreigners own at least 750,000 acres, or a little more than 1 percent of the total.

One of the most striking features of the changing pattern of agricultural landownership in Britain this century has been the entry into farming of landowners from different backgrounds and class, coupled with the move away from tenanted land to owner-occupation. During the First World War, the

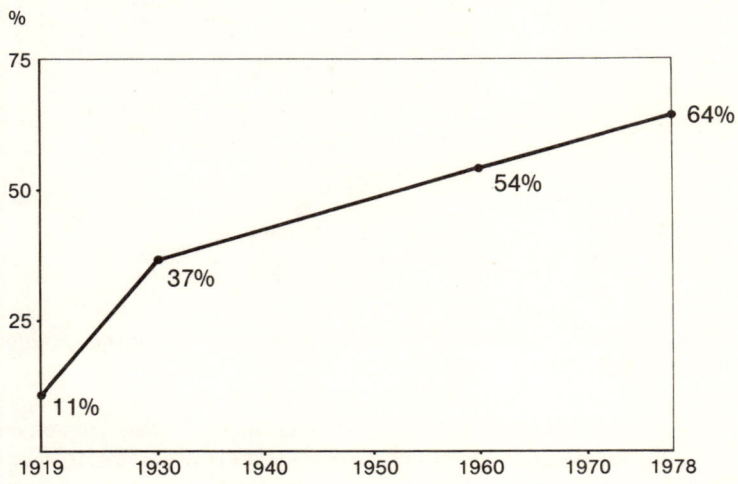

4. Owner Occupation: *the left hand scale shows the percentage of all farms.* *(Source: M.A.F.F.)*

sons of many wealthy landed families were killed and large agricultural estates, already threatened by the depression in farming over the previous fifty years, broke up. And many men who returned from the war on the continent became farmers for the first time.

The years following the war witnessed a change in land-ownership unprecedented since the dissolution of the monasteries in the sixteenth century. Land was cheap. About a quarter of the agricultural area of England changed hands in the two years of 1921 and 1922. A new breed of entrepreneurs — industrial capitalists — came onto the land market, attracted partly by the status it would give them. The other main group of buyers were the sitting tenants of landlords who were killed or gave up farming for the City.

There was another surge towards owner-occupation between 1950 and the early 1960s with some large landlords feeling the pinch of rising taxation and especially rising costs. Sitting tenants again accounted for a large share of the land purchases during this period. But this was also the time when a significant proportion of landlords began farming themselves land they had previously let, a trend that is still continuing.

Proportion of farmland rented:

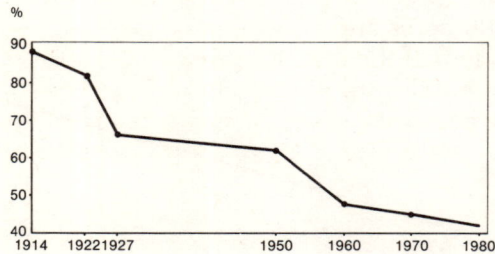

5. **Proportion of Farmland Rented:** *changes in ownership and taxation cause steep falls. (Source: M.A.F.F.)*

These are official figures, but because of the increased incidence of disguised ownership, to which we refer below, it is likely that much less land is rented our to tenants, perhaps only 35 percent.

The other striking trend is the movement towards fewer and

larger farms which, in turn, account for an ever-increasing proportion of Britain's food production. The number of small farms has been falling fastest while the number of farmholdings has been more than halved since the beginning of the century.

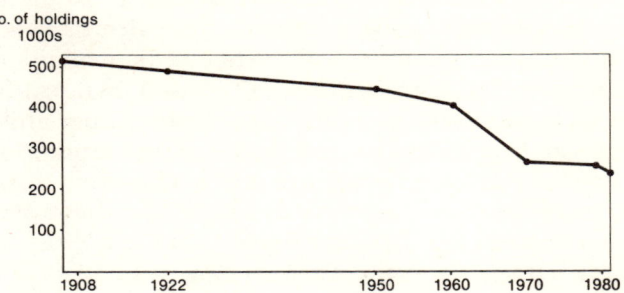

6. **Farmholdings:** *as the number of farms fall, size increases. (Source: M.A.F.F.).*

The average size of farm increased from 63 acres in 1908 to about 278 acres in 1980.

There are many medium-sized, family farms; about one-third of all agricultural land is accounted for by farms of between 300 acres and 1000 acres.

But nearly half the total number account for 90 percent of total agricultural output, and about 7 percent of the largest estates now account for half the total acreage — and the largest 10 percent, about 24,000 farms, produce half the food grown in Britain.

The private landowning sector is extremely difficult to unravel; it covers land placed in the hands of family companies, partnerships and trusts.

To avoid the impact of taxation, notably Capital Transfer Tax, and to maximise the benefits from grants schemes, the ownership of an increasing number of large estates with let land is being transferred to trusts, including charitable trusts. A recent Cambridge University study of holdings of more than 100 acres showed that trusts of various kinds 'owned' two-thirds of the area covered. As the Northfield committee put it: "it is clear that private individuals may own land through the medium of a variety of partnerships, trusts, and companies and that they may also own land solely or jointly with somebody else". Owners of large inherited estates — the Dukes of Northumberland and Devonshire, for example — can therefore claim that trusts and not themselves as individuals own the bulk of the land normally attributed to them.

Behind the fences and the hedges, tax advisers and accountants have created a labyrinth of trusts, companies and partnerships that lead to a great deal of disguised ownership of the larger farms and sabotage official statistics. Although farmers are required each year to give details of the amount of land they own or rent, the figures are misleading and the Government relies heavily on surveys conducted by the universities, the CLA (County Landowner's Association), the NFU (National Farmers' Union), and the estate agents.

Farmers sometimes set up trusts and companies which then legally own the land. The farmer can then 'rent' his 'own' land from his company or trust but tells the Ministry of Agriculture that the land is simply 'rented'. Research by David Rose and Peter Saunders at Essex University demonstrates that in this way official figures underestimate the number of farms of more than 1,000 acres in practice owned — as opposed to rented — by as much as half in Cambridgeshire, by a third in Norfolk, and by about one-fifth in Suffolk. They calculate, for example, that over 40 percent of the larger farms (over 1,000 acres in East Anglia) are companies.

So even though many of the biggest farms are legally owned by more than one person, in fact a very small percentage of landowners dominate the agricultural industry in the four counties of Essex, Cambridgeshire, Suffolk and Norfolk, which together contain no less than half of Britain's limited amount of Grade 1 farmland. In these four counties less than 10 percent of the total number of full-time landowners in England and Wales account for more than a quarter of the wheat acreage, a third of the total vegetable acreage, a quarter of the maincrop potato harvest and well over half the sugarbeet crop.

There are 43.8 million acres of agricultural land in Great Britain of which nearly 39 million acres, or about 90 percent, are owned by individuals, either directly or through trusts, family companies or partnerships. (See Fig 3, p28)

Over the country as a whole, about two-thirds of the estates of more than 10,000 acres (the most productive as well as the large parks) are still owned by individuals, again including trusts.

Going down the scale a little and taking another yardstick, 6 million acres in England and Wales are held in privately-owned estates of 5,000 acres or more (this figure has only halved over the past 100 years). So at the very most 1,200 individuals own a quarter of England and Wales. According to private estimates, about 1,500 individuals own about one-third of England and Wales.

The Country Landowners Association in England and Wales has about 45,000 members, but half of them own less than 100 acres. So about 23,000 members (though they include some institutions) own 90 percent of the land in England and Wales.

In Scotland, the top 100 landlords (including institutions) own about 5 million acres — about one quarter of the country — as against 10 million acres 100 years ago.

About 1,700 individuals own about a third of Britain; 100 years ago, about 7,000 individuals owned about three-quarters of the country. These must include most of the 300 members of the traditional landed aristocracy who have retained large tracts of land, and many of the 700 members of the traditional landed gentry who have large estates.

These figures offer some perspective to the official statistics which record that most farms are smallholdings and family farms. There remains an immense concentration of land-ownership in Britain, one that is unique in Europe.

With only between 1 percent and 2 percent of farmland changing hands each year, short of a radical and dramatic policy initiative, a change in the whole outlook towards agriculture and food production, it will be a very long time before the pattern and structure of landownership significantly shifts. And since traditional landowners, while losing their former significance and influence in agriculture are far from being relegated to history books, it seems appropriate to take a look first at the Monarchy.

The Royal Family

Under the feudal system the monarch owned all the land. His subjects occupied it and farmed it by permission of the King in return for service and money. The system was applied strictly by William the Conqueror and 900 years later land was still the most important source of royal patronage. Between the 12th century, their Golden Age, and the 16th century, the monasteries, with their highly disciplined society, remained wealthy and peaceful landowners. After the dissolution of the monasteries, Henry VIII distributed their property to families which supported him and to the Tudor dynasty. Elizabeth I, who was desperately in need of money to finance foreign expeditions and wars, took church land and sold it to her favourite courtiers and civil servants. Some monarchs became genuinely interested in agriculture and horticulture. James I planted a mulberrry orchard on the site now occupied by Buckingham Palace. But ironically it was George III, a dedicated agriculturist noted for the model farms he established in Windsor, who agreed, in 1760, to give up all rights to most of the Crown Lands, and, with them, the burden of paying for the running of the Government, in return for the Civil List voted by Parliament.

However, the monarch remains one of the largest private landowners. According to the author of the only comprehensive treatise on royal farmers: "In making the most of their rural inheritance . . . the Royal Family are doing exactly as most of the Queen's subjects would do if they could change places. Their attitude in this respect is typically British".[5]

The Queen's largest private property is the 80,000 acre Balmoral estate, to which she added 6,700 acres of grouse moor from the nearby Delnadamph estate for £750,000 in 1977. Buckingham Palace explained at the time that the Royal Family used to rent the moor, "but with rents going up all the time, it was better to buy one".

The other major private royal estate is Sandringham, Norfolk, with 20,000 acres of good farmland. The estate, which includes six villages, was bought by Queen Victoria for her eldest son, Edward VII. He, like his successors, was a keen farmer but his successors could not avoid controversy. In

1977, Jack Boddy, agricultural union organiser for Norfolk, claimed that the design and conditions of some of the cottages on the estate were below local authority standards. The Royal Estate was also slow to follow up changes in wages and rent legislation. When I asked him what were the wages at Sandringham, Boddy replied: "It is considered an honour to work there. The workers expect to receive their reward in heaven".

In common with many of her landowning subjects, the Queen over the past few years has taken direct control over land which was previously tenanted. She pays no tax on farm income from her private lands, though she does pay the local rates.

For Princess Anne and Mark Phillips, the Queen bought Gatcombe Park, a 750 acre estate in Gloucestershire, for an estimated £750,000 from the former Conservative politician, Lord 'Rab' Butler, who was Master of Trinity College, Cambridge, where Prince Charles was educated. Shortly afterwards she bought the nearby 530 acre Aston Farm which she is now leasing to her daughter and son-in-law. The Duke of Gloucester farms the 2,500 acre Royal estate at Barnwell in Northamptonshire.

The Duchy of Lancaster belongs to the monarch. Although the Queen does not personally own it, it was excluded from the 1760 Crown Estate-Civil List exchange. The revenue from the Duchy, in the order of £500,000 a year, goes to the monarch. The core of the estate, in Yorkshire, Lancashire and Staffordshire, including 20,000 acres of Yorkshire moorland, is the land which belonged to Simon de Montfort and Earl Ferrers before their defeat by the King in the thirteenth century. It includes 50,000 acres of rich farmland in the Fylde, property in the Strand and the City of London, and a large investment portfolio.

The Duchy of Cornwall, established about 750 years ago, is the personal property of the male heir to the throne. The estate includes 128,000 acres extending across six counties in the West Country, 4,000 acres on the Isles of Scilly, 160 miles of foreshore and 11,000 acres of river bed. His 45 acre estate south of the Thames in London, including the Oval cricket ground and the two-bedroomed flat in Kennington rented by former Labour Prime Minister James Callaghan, brings in almost as much income as the Duchy's farmland. The Duchy is also the largest landowner in Dartmoor — and as the property of the heir to the throne it is not bound by planning laws — with some 70,000 acres there.

The Duchy's total income amounts to some £500,000 a year. Following a tradition started by his uncle Edward VIII when he was Prince of Wales, Prince Charles paid half of the income (which is tax-free) to the Treasury. However, after his marriage in 1981 he reduced this to 25 percent of the total income. He also adopted a more aggressively commercial approach towards the management of the estate, dramatically increasing rents, in some cases by as much as 90 percent.

In 1981 the Queen bought Highgrove House in Gloucestershire with an estate of 347 acres from Maurice Macmillan, son of the former Conservative Prime Minister, for Prince Charles at a cost of between £750,000 and £1 million. The only land farmed by the Duchy of Cornwall's management is the 125 acre Home Farm at Stoke Climsland, Cornwall.

It is difficult to separate the Queen's private wealth from the wealth she enjoys in her role as the Monarch — the Royal stamp, antique and book collections, for instance. If the collections but not her overseas investments are included, the Queen's wealth was worth about £45 million in 1978. The Companies Act has been amended, following Royal pressure, to disguise just where the Queen's shares are invested. How much she accounts for of the £80 millions worth of shares kept secret in the thirty companies that make up the Financial Times index is not publicly known, though the Queen does have a substantial holding in Rio Tinto-Zinc, the large international mining company.

The Crown Estates

The Crown Estates are in essence the land given up to the Government by George III in return for the Civil List. They cover about 350,000 acres of farm and forest land, with some 200 tenants, and the revenue from them — about £6 millions a year, most of which comes from central London assets. They include part of the Romney Marsh in Kent, Tintern in Wales,

Drake's Island off Plymouth, and farmland in 25 countries of England. They include land reclaimed from the Wash, land in Richmond, Windsor and Ascot.

The Estates are administered by eight Crown Commissioners. Their domain includes 250,000 acres of agricultural land. The Commissioners are active on the farmland market, regularly buying and selling agricultural property. But the Estate's urban property in London is by far the most valuable and profitable. This includes parts of the City of London around Ludgate Circus, clubs in St. James's and Pall Mall, South Africa House in Trafalgar Square, the Strand Palace Hotel, the Haymarket, parts of Soho, offices in Piccadilly and Park Lane, much of Regent Street and the site of the Millbank Tower.

The Church

In 1066 the King, William the Conqueror, owned all the land. But it was not long before the Crown needed the Church as a source of money and as an instrument of patronage. English monarchs kept sees vacant to take the money otherwise due to the bishop in question. At the same time monarchs and barons gave the Church endowments, because they attached genuine importance to the Church's teaching.

Bishops and abbots managed large estates and in return respected their feudal obligations to the Crown. They were successful capitalists and key advisers to the King. They often presided over courts as Justices of the Peace, and often became the wealthiest landowners in the district. Monasteries grew wealthy, especially during the period of political troubles after Henry I's death in 1135. The Cistercians, the most enterprising order, sometimes ejected the inhabitants of entire villages to make way for their agricultural developments.[6]

So the Church, until the Dissolution of the Monasteries, was the richest, the most enterprising and, with land in every parish, became the largest landowner in Britain. But when its wealth and land was taken away, the Church, like the monarch, was forced to negotiate a close relationship with the nobility and gentry. As they left part of their land (glebe) or part of the produce of their land (tythe) to the Church, the Church in turn allowed lords of the manor to choose the local vicar. An estimated 2,000 church livings remain in private hands and many local landlords are still reserved the front pews in village churches. (Ministers in Scotland were directly supported by local landowners up to 1925.) The links between landownership and the Church were strengthened by the custom whereby the eldest son of the gentry, or even the aristocracy, would take over as lord of the manor from his father, the second would join the armed forces, while the youngest would be sent into the clergy. Today, the vicars in rural communities have a special status: they are respected pillars of the local community, frequently appointed to the agricultural housing tribunals which were set up after the abolition of the tied cottage system, while their wives are active in local charities and the Women's Institute.

In 1948, Queen Anne's Bounty, the system designed to help poor clergy and their churches, was merged with the Ecclesiastical Commission to become the Church Commission. With an unashamed secular bias, the Commissioners began to enter the stock market like any ordinary investor, though their payments are by statute restricted to churches, administration, salaries and pensions.

The Church of England is now the tenth single largest landowner, but although more than 90 percent of its 170,000 acres is farmland, it is also one of London's biggest private residential landowner, in 1980, after the Freshwater Group. Ninety-five Church Commissioners, including the Prime Minister, the Home Secretary and the Lord Mayor of London, are officially responsible for the property which covers 500 tenant farms concentrated around cathedral cities such as Canterbury, Chichester and Wells, and in the rich agricultural areas of East Anglia and the North-east. But closer control lies with the 30-strong Board of Governors, chaired by the Archbishop of Canterbury. The Board takes great pride in a convention that a Commissioner visits each farm personally at least once every five years.

Since the War, the Church of England, through its Estates Development and Improvement Company, has sold a great deal of low-rent housing to invest in the high-income, residential and commercial property market. It now has interests in the Barbican development in the City of London and, with London Merchant Securities, developed part of its crumbling Paddington estate. Though the Church still owns about 6,000 rented homes in London, including low-price, and in many cases, run-down, housing in Brixton, Lambeth, Hampstead and Chelsea, it has offered for sale all its old flats in its largest residential estate in the Little Venice — Sutherland Avenue area of Maida Vale. This was expected to bring in £25 millions for new investment.

But its policy towards investment and property, like the Crown's, has not escaped controversy. Over the past decade, the Church of England has declared redundant 840 churches and chapels, fewer than half of which have been put to new uses. In 1973, the Church Commission evicted squatters from empty houses in the Maida Vale estate and five years later tenants protested vigorously against a proposal to raise rents by 85 percent.

Income from the Commission's stock exchange holdings amount to above £20 millions and income from its property, a little less. Its assets total about £1,000 millions, with stocks and shares providing rather more of its investment portfolio than land and property interests. Farmland accounts for only about 6 percent of its total income.

Sir Ronald Harris, First Church Estates Commissioner in the late 1970's, reflected what has become a conventional City approach: "agricultural land has undoubtedly proved one of the best forms of investment from the point of view of both preserving and enhancing capital value and securing a reliable and growing income, and this must surely be the main incentive for an institution to own farmland. Indeed, it is arguable that any large and well-balanced investment portfolio must include farms".

This echoes precisely the view increasingly offered by financial institutions. Sir Ronald adds: "an underlying reason for the capital security of agricultural land is, of course, that it is one of the few 'real' forms of investment, a commodity of which there is in the United Kingdom a strictly limited supply and, indeed, what there is is steadily being eroded for developments of every kind".

A City businessman could not have put it better — land treated as a prime commodity, a capital asset producing secure profits. The Church Commissioners have shared the experience of other institutional investors in farmland, namely, that rental growth has increased faster than dividend growth from stock exchange shares. Over the past decade, the average rent per acre on Church farms rose by 208 percent; dividends over the same period rose by 71 percent.

Oxbridge Colleges

Oxbridge Colleges own about the same amount of farmland as the Church, some 160,000 acres, and their estates, which have diminished steadily since the War, have been a traditional source of income rather than a vehicle for increasing their investment portfolio. Most of the land is let out to tenants. Oxford accounts for two-thirds of the total. Their status as registered charities gives them tax privileges.

Christ Church, Oxford, tops the table with over 19,000 acres in eleven counties, followed by All Souls with nearly 16,000 acres. Trinity College, Cambridge, is known to have over 15,000 acres in East Anglia. For long it has been said that the traveller can go from Oxford to Cambridge without leaving land owned by Merton College, Oxford, and journey from Cambridge to London without leaving land owned by Trinity College. In addition St. John's College, Oxford, has important landholdings in London.

Other universities and independent schools own about 80,000 acres all together.[7]

Government Departments and Public Institutions

Central government departments own over one million acres of farmland, excluding forests and woods. By far the largest landowner in this category is the Ministry of Defence. Ever since village greens were used for archery practice, the land has been used for military training and exercises. The Defence Estate, which during the Second World War increased from 250,000 acres to 11.5 million acres or 20 percent of the total land area, now spreads over 700,000 acres (about the size of Northamptonshire), including 120,000 acres of foreshore, 198 miles (3 percent) of the coastline, 210,000 acres of airfields, barracks and depots.

The Ministry of Defence is also the largest owner of protected, or designated land — National Parks, Site of Special Scientific Interest and Areas of Outstanding Natural Beauty. About a quarter of its land lies in these areas and it has sites in all National Parks with the exception of Exmoor.

It lets about 130,000 acres of good farmland and about the same amount of poor grazing land to some 1,725 tenant farmers. The Ministry's holdings consist of 821 separate establishments ranging from single properties to large tracts of countryside — the largest is 92,000 acres of Salisbury Plain which makes the Ministry the largest landowner in Wiltshire.

In 1980 the Ministry earmarked 3,000 acres on the Plain adjoining the chemical defence establishment at Porton Down for chemical warfare battle training. Planning privileges — like other government departments, under circular 7/77, the Ministry is immune from public inquiries or objections — and the Official Secrets Act meant that the decision passed secretly and quickly through Salisbury District Council in the autumn of 1979. The Ministry had decided to transfer facilities to make riot control and toxic gases, previously produced on the Cornish coast at Nancekuke, to Porton. The move was not announced publicly even though a few years earlier that Nugent Committee reporting on defence lands stated: "It would not be practical to transfer the facilities to Porton because of, primarily, lack of space there, but also the problems of effluent disposal". The decision to go ahead and enlarge the Porton establishment will also increase water demands there by at least 120,000 gallons a day.

The Ministry owns 56,000 acres of firing ranges at Otterburn in the Northumberland National Park and has control over 34,500 acres on Dartmoor. The military use of Dartmoor is a source of perpetual controversy. The national park committee has invariably sided with the Ministry (which, in turn, enjoys the support of the Duchy of Cornwall which owns large acres of the northern and central parts of the Moor) against naturalist and amenity groups such as the Countryside Commission. One example involved the use of Cramber Tor,

Some Ministry of Defence Lands.

Dartmoor National Park: *rights of 34,500 acres for training and firing ranges.*
Peak District National Park: *training and firing range over 2,500 acres.*
Northumberland National Park: *training area over about 56,000 acres around Otterburn.*
North Yorkshire Moors National Park: *training area at Cropton Forest.*
Yorkshire Dales National Park: *barracks and training area at Catterick.*
Brecon Beacons National Park: *barracks and training areas 1,500 acres.*
Pembrokeshire Coast National Park: *training areas and ranges 30,000 acres, including 5,000 acres at Castlemartin.*
Snowdonia National Park: *airfield at Llanbedr and rifle range at Towyn 600 acres.*
Nancekuke, Cornwall: *former chemical weapons research establishment.*
Gruinard Island, Wester Ross: *contaminated by anthrax.*
Lulworth, Isle of Purbeck, Dorset: *firing ranges.*
Shoeburyness, Essex: *ranges, 30,000 acres (including foreshore).*
Holbeach, Lincs: *RAF weapons range.*
Longmoor, Hants: *former army camp and training area.*
Stanford, Norfolk: *army training area, 17,000 acres.*
Orfordness, Suffolk: *RAF and USAF radar station.*
Ashdown Forest, Sussex: *army training area.*
Chatham, Kent: *training area, 530 acres.*
Isle of Skye: *airfield being enlarged as a NATO base.*
South Uist: *army range.*
Kirkcudbright: *army establishment and ranges.*
Pembrey and Pen dine, Dyfed: *RAF weapons range and experimental establishment.*
Salisbury Plain, Wiltshire: *army ranges and training areas over about 90,000 acres.*
Porton Down, Wilts: *chemical warfare research establishment now being enlarged.*
Hamsterley Forest, Durham: *army training area, 2,000 acres.*
Ross-on-Wye, Herefordshire: *army rifle range.*
Malvern, Worcs: *army training area in a Site of Special Scientific Interest.*
Thetford, Norfolk: *army rifle range.*
Thorney Island, Sussex: *RAF airfield.*
Bentwaters, Lakenheath, Mildenhall, Woodbridge- Suffolk: *airfields used by the USAF.*
Molesworth, Cambs., and Greenham Common, Berks: *proposed sites for US cruise missiles.*

owned by the South-West Water Authority, for military training. Despite official policy that publicly-owned land in a National Park should be used for such purposes only if the need is demonstrated 'beyond doubt', the Dartmoor National Park Committee was quick to agree to the Ministry's plans for the Tor.

The 1973 report of the Defence Lands Committee, chaired by Lord Nugent, sympathised with the Army's overall case that it needed the land it had under its control. However, the Committee did propose that the army's gunnery school on the Isle of Purbeck in Dorset near Lulworth Cove should be moved to Castlemartin in Pembrokeshire where the Army was already using 6,000 acres on the coast. The plan was abandoned, partly because of protests from Welsh sheep farmers. In fact, Welsh farmers enjoy cheap grazing rights on the Army ranges during the winter. (Relations between the Army and farmers are less harmonious around the Army's 30,000 acres on the Eppynt mountain, north-west of Brecon. The Army on some occasions has tested long-range weapons, including the FH 70 howitzer, from Forestry Commission land rather than from its own ranges. "We've been firing over the top of civilian villages on Salisbury Plain for years without anything going wrong," the Army insists. "We're quite certain the FH 70 is foolproof.")[8]

Many of the inhabitants around Lulworth in Dorset wanted the Army to stay on, partly for economic reasons. But, the attitude also reflects traditional pride which the majority of the local population in rural areas share in the identification of their particular region with a specific government activity or service. The construction of the fast-breeder nuclear reactor at Dounreay, in Caithness, for instance, has given the place a new identity, as well as jobs. The Central Electricity Generating Board succeeded in persuading Suffolk County Council into believing that they were doing the country a service by planning to build Britain's first pressurised water nuclear reactor at Sizewell. East Anglia, of course, has also lived for a long time with British and American bomber bases.

But times are changing. Local comunities are beginning to question the assumption that Government agencies must be given priority over their own needs and interests. To the

Army's great embarrassment, fishermen off the Isle of Purbeck succeeded in 1978 in curtailing a demonstration planned to show off modern weapons to a group of visiting Chinese arms buyers. East Anglian representatives succeeded in winning overwhelming support for a notion condemning the siting of cruise missiles in Britain at the farmworker's union conference in 1980. Since then, opposition throughout the country to the presumption of Government and the nuclear industry that they can do what they think fit has grown dramatically. Decisions are still taken before any proper consultation with local inhabitants — the decision to expand the NATO airfield at Stornoway in the Hebrides, for instance, and to expand the nuclear submarine base at Coulport and Faslane on the Clyde, though opposition to them is now bubbling over the surface. Significantly, the main reason why the government, in June 1980, decided to stockpile American cruise missiles in two sites, Molesworth in Cambridgeshire and Greenham Common in Berkshire, against American advice (the U.S. wanted just one site) was an attempt by Whitehall to divide and disperse opposition.

The Ministry of Defence, with other government departments, has constructed a series of underground installations which are designed to be regional and military centres of government in times of tension, in what looks like a run-up to an outbreak of war, and during a war itself. The Ministry also controls a great deal of airspace, including virtually the whole of Wales, and it is increasing the area in Scotland over which it intends to practice low-level flying.

The Ministry of Agriculture owns about 35,000 acres, made up chiefly of experimental farms, Land Settlement Associations and the Farm Settlements estate.

The highly successful, but now neglected, Land Settlement Association estates were set up to provide smallholdings during the late 1930s to settle unemployed industrial workers on the land. The ten estates are grouped in Bedfordshire, Essex, Suffolk, Cambridgeshire, but there is also one in Newent, Gloucestershire, where tenants rebelled in 1971 against dictatorial management. The estates amount to about 4,000 acres in all. They are a good example of how tenant farmers *can* cooperate, by pooling their income from sales, coordinating their services and marketing — incidentally they provide the bulk of Britain's spring lettuce market. They actually make a profit.

The Farm Settlement Estate, set up to provide work on the land for soldiers returning from the First World War, now includes just two estates, in Sutton Bridge and Holbeach in Lincolnshire, with about 6,000 acres and 200 tenants. But they, too, have been neglected; they are loss-making and poorly-organised.

The village of Laxton in Nottinghamshire, the last surviving village using the open field, strip farming system — the basis of English agriculture and of village self-sufficiency for a thousand years — was sold by the Ministry of Agriculture to the Crown Estate in 1981 for £1 million. Fifteen farmers and half-a-dozen smallholders still farm the 1,800 acres and the traditional system is administered by the Court Leet of the Manor which consists of all the occupiers of land in Laxton. The bailiff, steward and a jury of twelve men have the power to inspect the fields and fine any tenant abusing the system.

Laxton was bought by the Ministry in 1952 when the former owner, the sixth Lord Manvers died, leaving a large bill for death duties. When it was put up for sale, widespread opposition persuaded Peter Walker, the Agriculture Minister, to give assurances that it would be sold only to a buyer who preserved the estate in its traditional form. The National Trust could not afford it but the Crown Estate eventually agreed to maintain it in its traditional form.

Other government departments which own large tracts of land include the Ministry of Transport which is also responsible for 150,000 acres of motorways and trunk roads. About 1¼ million acres of smaller roads are owned by local authorities. The Prisons Department of the Home Office owns 6,500 acres of farmland and 250 acres of forest. Water authorities, under pressure from increasingly thirsty and wasteful clients, own about 300,000 acres; the National Coal Board owns about 260,000 acres, of which half is let to farmers, and British Rail owns about 225,000 acres, much of it derelict.

The Forestry Commission, which ultimately comes under the control of the Department of Agriculture and Fisheries in Scotland, and the Ministry of Agriculture in England and

Wales, now has a little over three million acres at its disposal, more than two-thirds of which are under plantation. The rest is awaiting planting or is made up of agricultural and grazing land, or rock. 757,000 acres are in England, about 402,000 acres in Wales, and 1,980,000 acres in Scotland.

But the Thatcher administration's policy of hiving-off government assets also extended to land. Forestry Commission woodlands have been sold in deals the Government has done its best not to publicise. For example, the Commission has sold four square miles of Lochar Moss in Scotland. Since the Government's aim is to make money out of the deals, conservation agencies are unlikely to feature among the buyers.

The main conservation agency in England and Wales is the National Trust, which has about 400,000 acres, including 2,000 farms and 150 stately homes. It started up in 1895 with its first bequest — 4¼ acres of the Dinas Oleu cliffs overlooking Barmouth estuary in Wales. The Trust is on its way to securing one of its most ambitious plans: Operation Neptune and the ownership of 1,000 miles of coastline. By 1981 it had acquired about 500 miles. The Trust's grass-roots support has grown dramatically over the past few years and it now has over a million fee-paying members. But it is resented by many farmers and private landowners, partly on the grounds that as a charity, it enjoys tax-free status. *

The National Trust for Scotland owns about 82,000 acres. Its most significant recent purchase is that of the island of Iona, the birthplace of Christianity in northern Britain, from the trustees of the tenth Duke of Argyll. (The island was originally presented to the Dukes and the Clan Campbell as a reward for their part in the massacre of Glencoe.) The money was provided by a charitable foundation led by Sir Hugh Fraser, the stores tycoon. The price of £1.5 millions, though it includes an endowment for maintenance, compares un-favourably with the £110,000 estimated by valuers as the true

---

* The National Trust was involved in a major controversy in 1982 when its Council granted the Ministry of Defence a 99-year lease to build a military command bunker under the Trust's Bradenham estate in the Chilterns.

economic worth of the island and its fifty houses, thirteen crofts and two tenant farms; it is however, regarded as a place of spiritual energy and of pilgrimage.

The County Council Smallholdings Estate, about 378,000 acres owned by, or leased to, local authorities — but under threat of declining fast because of financial pressures on authorities — has traditionally been regarded as the first step on the farming ladder, enabling tenants to gain experience in farming with little or no capital before buying their own land. Its origins go back to the proposals put forward in the nineteenth century by Joseph Chamberlain, John Stuart Mill, John Bright and Jesse Collings at a time when England's traditional yeoman farming stock were being squeezed by cheap food imports and overwhelming priority was given to industry. In his Radical Programme of the 1880s, Chamberlain said that "besides the creation of smallholdings, local authorities should have compulsory powers to purchase land where necessary at a fair market price . . . to be let at fair rents to all labourers who might desire them, in plots up to an acre of arable or 3 or 4 acres of pasture".

The 1908 Smallholding Act empowered local authorities to acquire land. The scheme began well, and many holdings were allocated to returning soldiers after the First World War. But since 1945 the number of holdings has been halved and the average size, now about 36 acres, doubled. The demand grew — in 1979 Staffordshire was faced with sixty applicants for every vacancy it advertised, but the vast majority of the existing 8,000 or so tenants are unable or unwilling to afford to buy land of their own. The assumption that these county council smallholdings would provide a stepping stone into farming has thus lost all credibility. Though the smallholdings are profitable, when councils came under financial pressure from central government in the late 1970s they were tempted to sell off fixed assets in return for cash. More than 1,000 acres were sold off in 1977/78 alone. Lincolnshire decided to sell off 24,000 acres of rich farmland, worth about £50 millions, to pension funds and established neighbouring farmers when farms become vacant, and Derbyshire, Shropshire, Cambridgeshire, Norfolk and the West Midlands are all being tempted to sell off their land.

Lincolnshire has the largest area of county smallholdings, followed by Cambridgeshire and Norfolk.

About 40 percent of England and Wales is covered by areas with a special status, mainly National Parks or Areas of Outstanding Natural Beauty, both mainly in private ownership. In practice, as we shall see, their protection relies on a relationship of trust between amenity groups, private landowners and national or local authorities. Although it has generally worked well in the past, it is coming under increasing strain as pressure on the land, from both the conservationists' lobby and farmers, intensifies.

National Nature Reserves, promoted by the Nature Conservancy Council in cooperation with landowners, cover some 327,000 acres, access to much of which is restricted. One of the most recent reserves, announced in 1980 with the agreement of Lord Montagu, covers 1,600 acres between Beaulieu and the Solent. The Conservancy Council also designates Sites of Special Scientific Interest, many of which are on Ministry of Defence land, 'protected' from the public. The Royal Society for the Protection of Birds manages 78 reserves, covering 86,000 acres.

Common Land

There are between 1.1 million acres and 1.5 million acres of common land — the whole problem of registration prevents greater accuracy. Yet common land is not really 'common' if only because there is now no recognition in Britain of the concept of land without an owner. The public, as such, cannot own land, nor can an unidentified community. So there is no general right of public access to most of the commons, though there are a few exceptions including the heart of the Lake District National Park, Hampstead Heath, parts of Dartmoor and the Welsh Mountains. Hundreds of village and town greens are also open to the public under special legislation.

The law is confusing. The Law of Property Act of 1925 gives the owner of common land the right to grant public access by means of a deed. It also protects commons against enclosure and encroachment.

Most common land in England and Wales — there is none

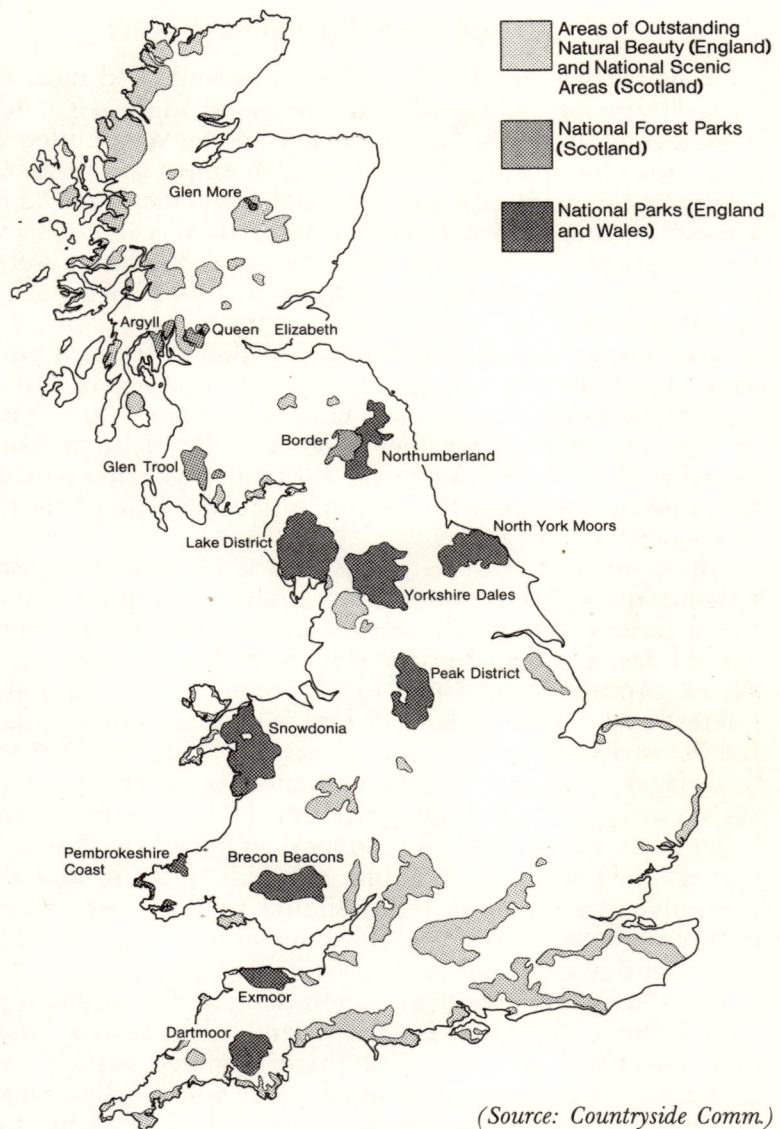

Areas of Outstanding Natural Beauty (England) and National Scenic Areas (Scotland)

National Forest Parks (Scotland)

National Parks (England and Wales)

Glen More

Argyll Queen Elizabeth

Border

Northumberland

Glen Trool

Lake District

North York Moors

Yorkshire Dales

Peak District

Snowdonia

Pembrokeshire Coast

Brecon Beacons

Exmoor

Dartmoor

(Source: Countryside Comm.)

7. National Parks and Scenic Areas: *nearly 30% of all Britain, including Areas of Special Scientific Interest, are "officially designated". But they are increasingly under threat from commercial pressures.*

so designated in Scotland — is in the uplands and most is owned by private individuals. Common land originated in the Middle Ages. It was the land of the Lord of the Manor offered as a source for grazing animals and gathering wood for fuel. The privileges became enforceable rights and the rough land known as 'common' because the rights were exercised by several people in common. Most of these commons were divided up into small fields during the Enclosures of the 18th and 19th centuries.

The rights of common traditionally included: the common right of pasture — the right to graze animals; estovers — the right to take small branches and bracken; turbary — the right to dig turf or peat for fuel; piscary — the right to fish; pannage — the right to allow pigs to eat acorns or beech mast fallen to the ground; and common in the soil — the right to take stone, sand or gravel.

After years of misuse, neglect, lack of consistency in policing these rights, which were eroded or forgotten, the Royal Commission on Common Land in 1958 declared that "as the last reserve of uncommitted land in England and Wales, common land ought to be preserved in the public interest". All common land, it said, should be open to the public "as of right" subject only to reasonable bye-laws. It said that a legal public right should be granted by statute subject to such restrictions as depositing rubbish and lighting fires, that wider powers should be given to local authorities to draw up management schemes including tree planting, and that all commons and common rights should be registered. Over twenty years later, only one of the recommendations — the registration of common land — had been enacted.

The 1965 Commons Registration Act gave five years for a kind of Domesday Book of common land to be drawn up. But little has been achieved. Just three commissioners were appointed to hear the many hundreds of disputed registrations. Witnesses to the true status of the commons had either died or moved elsewhere, with the case for commons lost by default forever. As the Commons, Open Spaces and Footpaths Preservation Society has pointed out, land which should be preserved as commons has been enclosed or cultivated.

According to the Society, the main problems are the

absence of a legal right of access for the public and the lack of adequate powers of management. It has proposed management plans drawn up by owners, commoners and local authorities. A few management schemes have been set up for regulating grazing and access on the basis of yet another law — the 1899 Commons Act — but, as the Countryside Commission pointed out in a paper on uplands published in 1978, they are virtually useless when the ownership is so uncertain.

The problem was clearly demonstrated in 1976 when the High Court deregistered common land on the Dee Marsh saltings in North Wales in favour of the Central Electricity Generating Board, which wanted to build a nuclear power station there, and against Clwyd County Council. It said that no specific commons rights had been registered.

The proposal by the Royal Commission (supported in 1978 by a Whitehall working party) for a universal right of public access to commons has been bitterly attacked by the National Farmers' Union and the Royal Institution of Chartered Surveyors. "It should be remembered", insists the NFU, "that common land is in fact private land, and that rights of common enjoyed on it may be vital for the viability of many farms".

"The rights," echoed the RICS, "are not held by the community, but by individual farmers and landowners".

The Aristocracy

> "The break-up of large estates is inevitable" – *the Duke of Northumberland.*
> "If we are still here, it is because you want us. If you did not, we would have been got rid of' – *Gerald Grosvenor, sixth Duke of Westminster.*

The British share a unique and remarkable respect for the hereditary principle and a fascination for the aristocracy. The aristocracy earns sympathy when it suffers from financial pressures. We applaud its expensive eccentricities and many of us seem to accept that it should be shielded from ordinary material problems. It is as if we would be lost without it. In turn, the aristocracy claims, with much truth, that it cares in its landowning role for the long-term interests of the country- side, that it preserves large sweeps of parkland, many stately homes and other, less tangible, reminders of Britain's history. And behind this, lies the aristocracy's genius for self- preservation.

Twenty-six dukes own something over 1 million acres and two-thirds of the landed nobility, about 200 families, own at least 5,000 acres each. About a third of the nine hundred or so hereditary peers are extensive landowners and the 'old titles' still account for the bulk of the largest private estates. The following table shows holdings in England and Wales (see p.00 for Scotland).

The Duke of Northumberland 90,786 acres
The Earl of Lonsdale 69,000 acres (Westmoreland and Cumberland)
The Duke of Devonshire 62,300 acres
The Duke of Beaufort 52,000 acres (Gloucestershire)
Lord Feversham 45,600 acres (Yorkshire)
The Duke of Westminster 37,000 acres (Cheshire, North Wales, Shropshire)
The Marquis of Exeter 26,000 acres (Lincolnshire)
Lord Egremont 20,000 acres (Sussex and Cumbria)
The Duke of Norfolk 18,400 acres
The Duke of Rutland 18,000 acres
The Earl of Halifax 18,000 acres
The Marquis of Bristol 16,000 acres (Suffolk)
The Duke of Grafton 10,250 acres (Norfolk)
The Marquis of Northampton 10,000 acres
Viscount Cowdray 10,000 acres (Sussex)
The Duke of Newcastle 9,000 acres (Dorset)
The Duke of Wellington 8,650 acres (plus estates near Waterloo in Belgium bequeathed to the first Duke)
The Duke of Bedford 8,400 acres
The Marquis of Abergavenny 7,500 acres
Baron Dulverton 5,500 acres (head of the Wills tobacco family)
Baron Vestey 4,600 acres (the meat and shipping magnate)

Sources: author's inquiries, with additional figures from R. Perrott, The Aristocrats, London 1968.

A random sample of twenty-five hereditary families shows that their estates declined from a total of 1.1 million acres in 1873 to 476,000 acres a hundred years later. They include the Duke of Northumberland, whose family estate fell from 186,300 acres, and the Duke of Devonshire, whose family estate declined from a total of 138,500 acres then. Both Dukes believe that the days of large private estates are numbered.

A few, including the Duke of Beaufort, the Earl of Lonsdale and Lord Feversham, have increased their acreage; one or two, like the Duke of Newcastle, have succeeded in rebuilding their ancestors' estates. Equally, a number of industrialists, more recently ennobled, have become large landowners.

Lord Leverhulme, the soap tycoon who was given a Viscountcy in 1922, owns 90,000 acres in Cheshire. The Earl of Iveagh, of the Guinness family ennobled in 1919, owns 24,000 acres in Norfolk. The Wills family is the second largest landowner in Scotland. Lord Cowdray, head of the Pearson financial and publishing empire and one of the richest men in England, besides his estate and polo park in Sussex, owns 88,000 acres in Scotland and a stretch of salmon fishing on the river Dee. His son, Michael Pearson is a large landowner and his estate includes 4,500 acres in Gloucestershire. The Vestey family owns 4,000 acres in the Cotswolds, 93,000 acres in the North-west Highlands, 2,000 square miles in Brazil, a further 1,000 square miles in Venzuela and leases 20,000 square miles from the Australian government in addition to landed interests in New Zealand.

Many peers, of course, have substantial investments and business interests to back up, and frequently help pay for, their country estates. Lord Inchcape, for instance, is director of more than 20 companies, including BP and Guardian Royal Exchange, as well as chairman of P & O, the shipping company founded by his father. The family also owns an estate of 13,000 acres, a 14,000 acre deer forest in Scotland, and a pheasant shoot in Essex.

Some have maintained their urban properties. The Cadogan Estate still covers about 90 acres of land in Chelsea, now under the control of Viscount Chelsea, son of the Earl of Cadogan. The Duke of Portland's family, including Lord Howard de Walden, owns over 100 acres in the Harley Street area north of

Oxford Street, as well as 17,000 acres in Nottinghamshire. The Duke of Portman's family, the Lyttletons, owns 55 acres around Portman Square and Baker Street. The Duke of Bedford owns about 30 acres in London. The Duke of Devonshire, whose family has contributed a great deal to the development of Eastbourne, including the building of schools and hospitals, continues to benefit through revenue from extensive property interests there. The Duke of Norfolk has substantial property interests in Sheffield as does Lord Calthorpe in south Birmingham. (The identification of individual families with particular cities continues and offers another interesting field of study. The Colman (mustard) family, for instance, who have been benefactors and patrons in Norwich, also have large commercial interests there, including control over the Eastern Daily Press.)

But perhaps the best example of the ability and determination of a family to hold onto its property is the Grosvenors, the family of the Duke of Westminster, a title granted only recently — by Queen Victoria in 1874 — in deference, it is said, and admiration of the family's enormous wealth. It is the last dukedom created in Britain. The family's fortune is estimated at between £300 millions and £1,000 millions. The jewel of the estate is 300 acres of Belgravia and Mayfair.

These valuable parts of London W1 and SW1 came to the family in 1667 when Sir Thomas Grosvenor married Mary Davies, a twelve year-old heiress. Her dowry was the Manor of Ebury, then mainly marshland and undrained bog. It now includes Eaton Square, Chester Square and property divided in equal proportion between diplomatic, commercial and residential interests. It covers Claridge's and the United States embassy in Grosvenor Square, the only American embassy in the world not owned by the US government. When Washington tried to buy the freehold, the Duke of Westminster offered to swap it for 12,000 acres of east Florida confiscated by the American government after the War of Independence. Washington chose instead a 999-year lease, with £1 annual ground rent on its London embassy.

The family, not surprisingly, has had to face the problem of death duties. It sold its Pimlico estates to help pay £12 millions in duty in 1953. However, the family fortune was immediately placed in a twenty-part trust fund, and when the sixth Duke succeeded to the title in 1979, he did not have to pay a penny of capital transfer tax. (A year earlier, the High Court came up with an added bonus by deciding that no death duties need be paid on the fourth Duke's £4 millions personal estate on the grounds that his death in 1967 by blood poisoning was the result of a war wound inflicted while he was commanding a tank regiment in France in 1944.)

The estate has thus been allowed to expand and now includes a 10,000 acre sheep farm in Wagga Wagga, Australia, and property developments in Vancouver and Hawaii.

It is probably not an exaggeration to say that most titled landowners, indeed the great majority, own assets, if not enjoy annual incomes, worth millions of pounds. What they provide — and believe they alone can provide — is well illustrated in a letter the Duke of Buccleuch, Britain's largest private landowner with extensive estates in Scotland and Northamptonshire, sent to Lord Northfield during the inquiry into changing patterns of farmland ownership.

Last century, he said, his family owned more than half a million acres, but as a result of rising costs and estate duty, the holdings had been reduced by about a half. Yet over the past twenty years, the family had contributed more than £1.75 millions worth of new capital for farm improvements. The Buccleuch estates had promoted a great deal of pioneering work, built cottages, churches, roads, bridges, erected fences, planted woodland, helped research institutes, encouraged young farmers by providing working capital. It had opened great houses to the public.

"Almost every village and town with which we have been connected", the Duke wrote " has its church, hospital, park, playground, cricket field, tennis court or hall given or con-tributed to by my family . . . local angling clubs have miles of trout and salmon fishings on all the major rivers and streams for nominal rents." "The advantage of large estates over small, he added, "is that different interests can be planned — farming, wildlife, amenity and landscape. They have better resources to help tenants, train specialists and good calibre managers, use machinery and adopt a flexible attitude to the question of public access."

Fifty years ago, the estates were turned into a limited company with shares held mainly by family settlements through discretionary trusts. But the Duke said that there was "little doubt" that they would continue to be reduced as a result of Capital Transfer Tax and Capital Gains Tax. He believes that increasing economic pressure on farmers is forcing them to concentrate on quick-yielding crops at the expense of both long-term investment in trees, for example, public access and recreation. "The countryside," he wrote in the *Sunday Times* in 1977, "is at risk as never before".

There is no doubt that there is great financial pressure on some members of the landed aristocracy. Patrick Montague-Smith, editor of *Debrett's Peerage and Baronetage*, wrote in the 1976 edition: "those who have two addresses have often deleted the London house or flat": they were feeling the pinch and returning to the hills.

The Duke of Northumberland makes the important point that capital for large landed estates in the nineteenth and much of the eighteenth century came not from agriculture, but from coal and industry. Equally, however, there are many titled and landed families which have substantial business interests in the City or, through a combination of luck and foresight, have set up trusts which shield them from death duties.

The appeal to many people of the aristocracy is that it is not made up of 'hard-headed' investment farmers and that it is, indeed, concerned about conserving the countryside. It also remains the bedrock of the landlord-tenant system — something which may explain the fact that the Duke of Buccleuch, Britain's largest private landowner is chairman of the Small-farmers' Association. This approach is shared by many of the gentry — the 2,000 families, named in *Burke's Landed Gentry*, which have owned 300 acres or more, and half of them at least 1,000 acres, for several generations. The difference is that few of these families have financial interests outside farming. And the gentry, too, is losing some of its land; a survey in the Country Landowners' Association magazine in the spring of 1980 suggested that only 15 percent of families owning an estate 100 years ago still maintained them. An additional 400 or so families which had acquired them since 1914 were holding on to them.

If the countryside is at risk, so, too, are country houses. According to Save Britain's Heritage, the organisation devoted to protecting historic buildings, more than 600 country houses were demolished between 1945 and 1974.

Successive governments have adopted a complacent attitude towards Britain's heritage, allowing half-baked tax measures, designed mainly as a superficial sop to sections of political opinion, to take their course without placing something else in

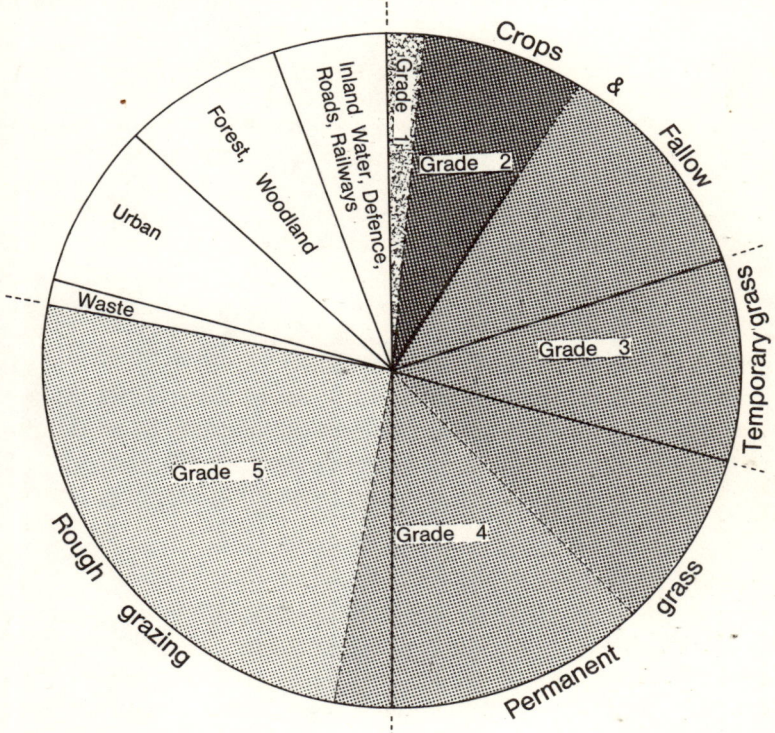

8. **Grades of Land:** *the circle represents all land in Great Britain. Most agricultural land in Britain is under grass; nearly a third is moorland or used for rough grazing. Yet a lot of grassland is badly or under used, and neglected. The official grading of farmland by the Ministry of Agriculture is misleading and confusing. In particular, the Grade 5 land is not being used to its potential. While wasteland and derelict land is increasing, industry continues to be allowed to take the easy, and cheaper, option – developing green field sites. (Source: M.A.F.F.).*

their stead. We have seen how large private estates can remain intact, but if some traditional landed estates are to die, why allow them to die slowly? Who should take them over? Can others save the countryside? Should these estates be parcelled out to new farmers?

In the meantime, let us look at the relative newcomers to the farmland market.

2

THE NEW DUKES

"A gentry only now and then residing at all, having no relish for country delights, foreign in their manners, distant and haughty in their behaviour, looking to the soil only for its rents, viewing it as a mere objection of speculation, unacquainted with its cultivators, despising them and their pursuits and relying for influence not upon the goodwill of the vicinage but upon the dread of their power." – *William Cobbett.*

Cobbett's distaste for the nineteenth century gentry is reflected today in concern about the role of insurance companies and pension funds. With the help of enormous fiscal and financial advantages, they are becoming the new dukes, not least by taking on the mantle of farming landlords as more and more private landowners are getting rid of tenants at the earliest possibility to farm their land themselves.

Pension fund managers enjoy enormous power and influence. The former Labour prime minister, Sir Harold Wilson, discovered while chairing a committee investigating the City, that "pension funds could well be transforming the nature of our society more than any government would ever dare to do even if it had a large majority in Parliament". Almost unnoticed, over the past fifteen years, the savings institutions have increased their ownership of British industry to more than 50 percent of all quoted stocks and shares. They control about £50,000 millions worth of shares and have concentrated on shares in large companies which account for 80 percent of stock market capitalisation. The pension funds alone control at least £25,000 millions of assets, a sum equal to twenty percent of the country's Gross National Product in

1979. In 1977, they invested over £500 millions in property alone. What *The Economist* has termed "a private corporate state" is being created, and while in theory these staggering amounts of money belong to millions of workers, in practice it is the fund managers, like unaccountable bureaucrats, who are controlling what amounts to the single most important source of finance for British industry and commerce. With the combined assets of pension funds and insurance companies growing at an estimated £3,000 millions annually, they are able to dictate their terms, including interest rate levels, to governments dependent on borrowing. Pension funds sometimes bale out their own, parent, companies.

These financial institutions are sponsors of sports and, sometimes, patrons of the arts. They also invest in art; the British Rail pension fund has bought at least £23 millions worth of antiques and works of art. They have invested in valuable property; the BP pension fund bought the Berkeley Estate in London. Given the amount of money at their disposal it is not surprising that they have also invested in farmland. The amount of new money available for investment by financial institutions is thirty times that required to purchase every single acre of land coming onto the market.

City institutions pounced on the farmland market in the early 1970s. In one year, 1972, they increased their agricultural holdings by 40 percent. Between 1971 and 1978 the City interest in farmland increased from 50,000 acres to at least 530,000 acres. The Northfield committee on farmland acquisition forecast that by the year 2020, they would own 4.75 million acres, or 11 percent of the total agricultural area.

Less than 2 percent of Britain's farmland goes onto the open market each year and available records show that in any one month purchases by financial institutions rarely account for less than half of all sales.

But the Centre for Agricultural Strategy at Reading University has forecast that if the institutions made a concerted effort and bought up most of the large farms coming up for sale, they could control about half of Britain's total food production within thirty years. The Committee of Property Unit Trusts, which groups over a thousand pension funds, told the Northfield inquiry that "it is increasingly likely that the

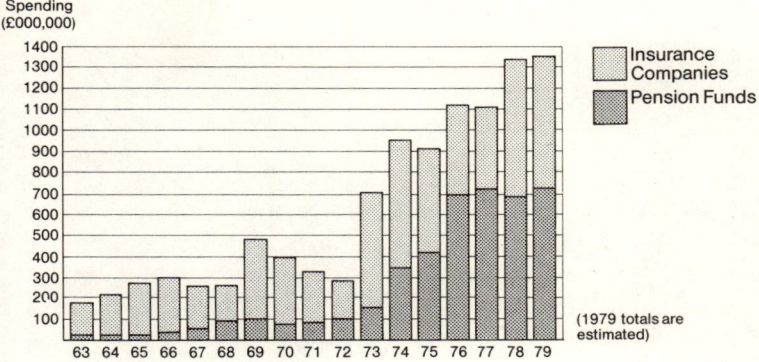

9. Institutional Investment in Land: *the financial power of insurance companies and pension funds – the growth of what has been called the "private corporate state" – is the cause of increasing concern. The amount of money available for investment by financial institutions is thirty times that required to purchase every single acre of land coming on the market. (Source: Strutt & Parker).*

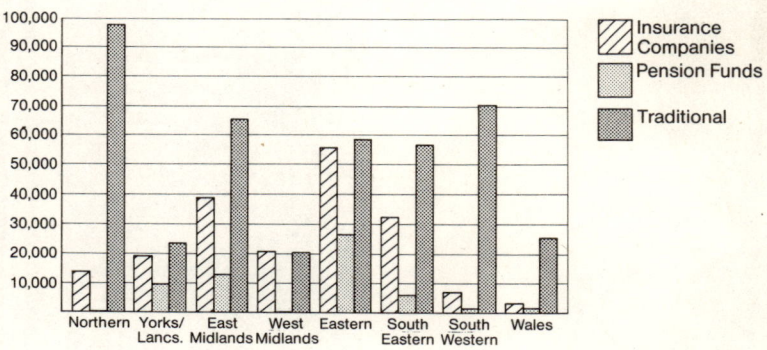

10. Institutional Ownership of Land by Region: *financial institutions concentrated first on the rich farmland of East Anglia which accounts for half of all Grade 1 land and where less than 10% of the total number of farmlandowners in England and Wales account for more than a quarter of the total wheat acreage. But the insurance companies and pension funds are now looking further afield – to dairy farming in Southern England, for instance, mixed farming in Wales and forestry in Scotland. (Source: E.D.C. for Agric.).*

Harvesting in Lincolnshire. *With the trend towards more specialised agricultural systems and away from mixed farming the East has become increasingly dominated by arable products. In some areas 80% of all trees have been cut down in the past twenty-five years and thousands of miles of hedgerows uprooted.* (Photo: Don McPhee)

institutions will become the main providers of let land . . . (the trusts) are almost alone in being in a position to provide the modern fixed equipment so essential to increase the industry's productivity and profitability". Moreover, the bare statistics revealing how much farmland they own hide the fact that the institutions have concentrated on good quality arable land, mainly in East Anglia and Lincolnshire, on which a significant part of Britain's food production depends. This is the land sought after as "prime investment" by the City. More recently, they have been prepared to seek out the best estates going on the market as far afield as Wales, Gloucestershire and Lancashire.

Some examples illustrate the trend:

Lord Wimborne sold 2,850 acres of his Ashby St Ledgers estate, including the stately home, in May 1976 to the British Airways pension fund for £2 millions. The village pub, the local inhabitants were assured, would be protected. Wimborne also sold part of the family estate near Poole to the local authority for £7.5 millions.

The Electricity Board superannuation fund bought 35,000 acres in Carmarthenshire from Lord Cawdor in August 1976, for about £4 millions. The seventy estate farmers had earlier demanded a meeting with the Earl to seek first option to buy the land. The tenants were told not to worry; their jobs would be guaranteed. But the rents were tripled within three months of the new owners taking over.

Lord Howard de Walden sold 8,000 acres including 48 tenanted farms and 450 acres of woodland in Ayrshire in 1978 for about £4 millions to the Pension Fund Property Unit Trust which by then had already invested about 20 percent of its portfolio in farmland, some 56,000 acres in all. The British Gas pension fund bought 7,500 acres of farmland between 1977 and 1979, keeping about half in-hand. The remainder it let to tenants.

In the course of one week in the summer of 1977, Gallaher's pension fund spent £5 millions buying nearly 3,000 acres of farmland, including cottages and woods, in Lincolnshire. The fund managers said that they were not interested in letting to tenants. They would farm the land themselves in partnership with Velcourt Farmers Ltd., a company which manages 20,000 acres of estates for both financial institutions and individual absentee landlords.

The Post Office staff superannuation fund in 1978 bought the Newbold Revel estate in Warwickshire, listed in the Domesday Book and at one time owned by Sir Thomas Malory, author of *Morte d'Arthur*, complete with 8 cottages and houses, for use as a middle management training college. The Post Office pension fund, the largest in Britain, had by then purchased 24,000 acres of farmland and intends to buy more because, as the managers put it, it is investment in one of Britain's most efficient industries and vital to the nation.

Hill Samuel's Mutual Agricultural Property Funds by 1979 owned 6,000 acres of good quality land in East Anglia which it lets to Hillcourt Farms Ltd. to manage jointly with Velcourt. A fund spokesman said at the time: "an efficient farm is of greater benefit to the community than a badly run one and we intend to be good neighbours".

Kleinwort Benson's Farmland Trust owns over 8,000 acres in East Anglia. The 7,000 acre Norton estate near Lincoln, once owned by the former prime minister, "Prosperity Robinson", who died as the Earl of Ripon in 1859 — was sold in 1975 by the Smith's food group to British Field Products, a subsidiary of Guardian Royal Exchange Assurance.

In 1976, nine farms near Beverley on Humberside were sold to the Alliance Insurance Company for more than £1 million. In February 1978, Captain Bulver-Long complained at a meeting of the Nafolle branch of the Country Landowners Association that the last three land acquisitions in the area had been by "a foreign bank, a pension fund and a large company". A few months later, Equity and Law Life Assurance Company spent £5.5 millions on farmland in Lincolnshire and since it was interested only in vacant possession sold those parts of the estates that were tenanted to another insurance company, UK Provident.

The most dramatic purchase of agricultural land by a financial institution came in 1979 when the Prudential Assurance Company spent £20 millions buying the late Sir Charles Clore's 16,600 acre Guys estate in Herefordshire — probably the largest farmland investment ever to have changed hands in Britain in recent times. Sir Charles, who bought the land from

Guys Hospital in 1961 for a reported £1.25 millions, had said that he wanted the new owner to keep the estate intact — and, indeed, there were few individuals who could have afforded to buy the whole of the estate. The Prudential's previous largest acquisition was the 8,000 acre South Esk estate in North-east Scotland, but the company has refused to disclose its total farmland investment.

Other insurance companies owning farmland include Eagle Star, Norwich Union, Legal and General, Commercial Union and Pearl. Pension funds of individual companies, Boots and Pilkingtons, for example, as well as those of nationalised industries, have invested in agricultural land. Companies dealing in farmland unit trusts and bonds include the Barclays Bank Trust, Matthews Wrightson Land and the Farmland Trust Company set up by Save and Prosper. The trend continues: the August 21 1980 edition of *Big Farm Weekly*, for instance, described transactions worth £9 millions covering ten farms and 8,000 acres handled over the previous month by the estate agents, Knight, Frank and Rutley. Of these two at least, a grade 1 fen farm in Cambridgeshire (Waldersley Farm, Wisbech), and a 1,100 acre arable farm in Gloucestershire, were bought by institutions. At the same time, food companies are buying their own farms: Sainsbury, for instance, has its own pig farms and a beef farm in Ayrshire. Dalgety own pig farms in East Anglia.

The ownership of farmland by the traditional institutions — the Crown, the Church of England, the Oxbridge Colleges — poses no threat to the status quo and is accepted by private landlords. But the emergence in the shires during the 1970s of financial institutions was as provocative to individual farmers as it was welcomed by the big estate agents which have profited from this new breed of wealthy clients. 'City money' entering a preserve known for its fierce independence, hostile to inter-ference from London, is resented by private landowners struggling to join the fight for an increasingly competitive and capital dependent agriculture. It is also resented because financial institutions are immortal in law; that is to say, they are not subject to capital transfer tax.

The traditional landed classes argue that the financial institutions cannot have the same sympathy towards rural life.

The Duke of Westminster claims that he is a better landowner than 'faceless people' in the City. "It must be accepted," says Roger Paul, former president of the Country Landowners Association, "that the essence of the difference between the institutional and the private landlord is that the former is accountable to his board, the latter to his conscience, his family and the community". Such is the antipathy, rarely expressed in public, between the old and the new landowners that the CLA refused to accept an advertisement in its house magazine from the National Union of Mineworkers' pension fund, a member of the Association in its own right.

The City's invasion of East Anglia has led farmers there to propose a limit, of perhaps 1,000 acres, on the amount of land any one individual or company should own. Meanwhile, private landlords say that the ownership of farmland by insurance companies could be a step towards back-door nationalisation of land, given the Labour Party's threat to nationalise these companies.

For their part, financial institutions protest that they channel only a small percentage of their total holdings into farmland — 7 percent in the case of Abbey Life, 5 percent in the case of the Post Office pension fund, and a similar proportion for the Prudential — in property as a whole. But this is small comfort compared to the total size of their assets. They claim that they will care for the countryside as much as their private predecessors. Ironically, senior Labour politicians have justified their involvement in farming by arguing that they are more democratic than the traditional private landlord-tenant system; since the financial institutions represent thousands of small investors, they say that their activities should be regarded as a an opening for the urban majority allowing it to take a stake in the countryside. Spokesmen for the City of Westminster's Farmland Fund, one of only two funds which are open to individual investors, claim that "the rolling and highly productive agricultural land of the Eastern counties, traditionally the preserve of the very rich, is now an investment within the range of every modest investor who wants to share in one of the nation's most valuable assets, food-producing land, because the company has pitched the minimum investment at £200". The Duke of Northumberland dismisses these arguments by saying they are as valid as the suggestion that the coal and steel industries are owned by the people because they are nationalised. He feels threatened by the growing power of the financial institutions. This conflict, over capitalism and landownership, is certain to take on a much more pressing and wider significance.

One of the most articulate and aggressive defenders of the entry into agriculture of large finance houses is Tony Rosen, former managing director of Fountain Farming, one of the first companies to offer to manage land in partnership with individual investors and shareholders in the company's parent, the Matthews Wrightson group. In its heyday, the group controlled about 45,000 acres of farmland and 120,000 acres of forestry.

"Nepotism," says Rosen, "is rampant in agriculture and most farmers think they are immortal. We all know that the best qualification to be a farmer is not necessarily to be a farmer's son. In too many cases, the main criterion of the owner-occupier is that his net spendable income should be enough to keep him and his family in the style to which they have become accustomed". Hereditary farmers and good farmers are seldom synonymous, he argues. And if private landlords cannot now afford the capital expenditure necessary to maintain 'an efficient business', why should the tenant be obliged to pay out?

Fountain Farming said in 1978 that not only was the company employing more people on its farms than before it took them over, but it had also planted more trees and hedges and built more cottages. The financial institutions insist that only they can provide the necessary capital and career structure that agriculture will need in the future. They say they invest more per acre than private landlords, who are placing an increasingly heavy burden on their tenants, and very much more than traditional public landowners whose record is the worst of all. Fountain Farming collapsed. "In the case of many of the larger farms the sums required for working capital have now reached levels which are beyond the average farmer's ability to provide," one pension fund manager said in 1978. That this is a simplistic conclusion, we shall observe later.

One formula the institutions offer is a sale and lease-back

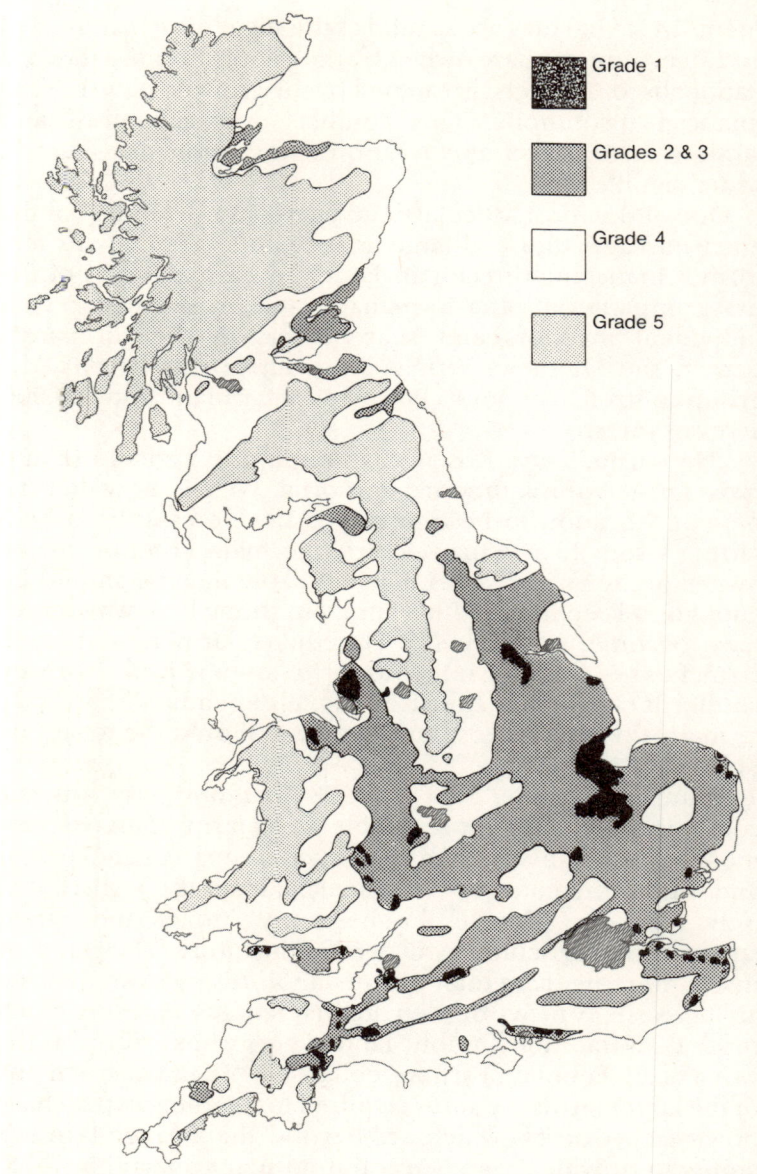

Grade 1

Grades 2 & 3

Grade 4

Grade 5

11. **Quality of Farmland:** *note that Grade 1 land is mostly near urban areas (except the Fenland). (Source: M.A.F.F.).*

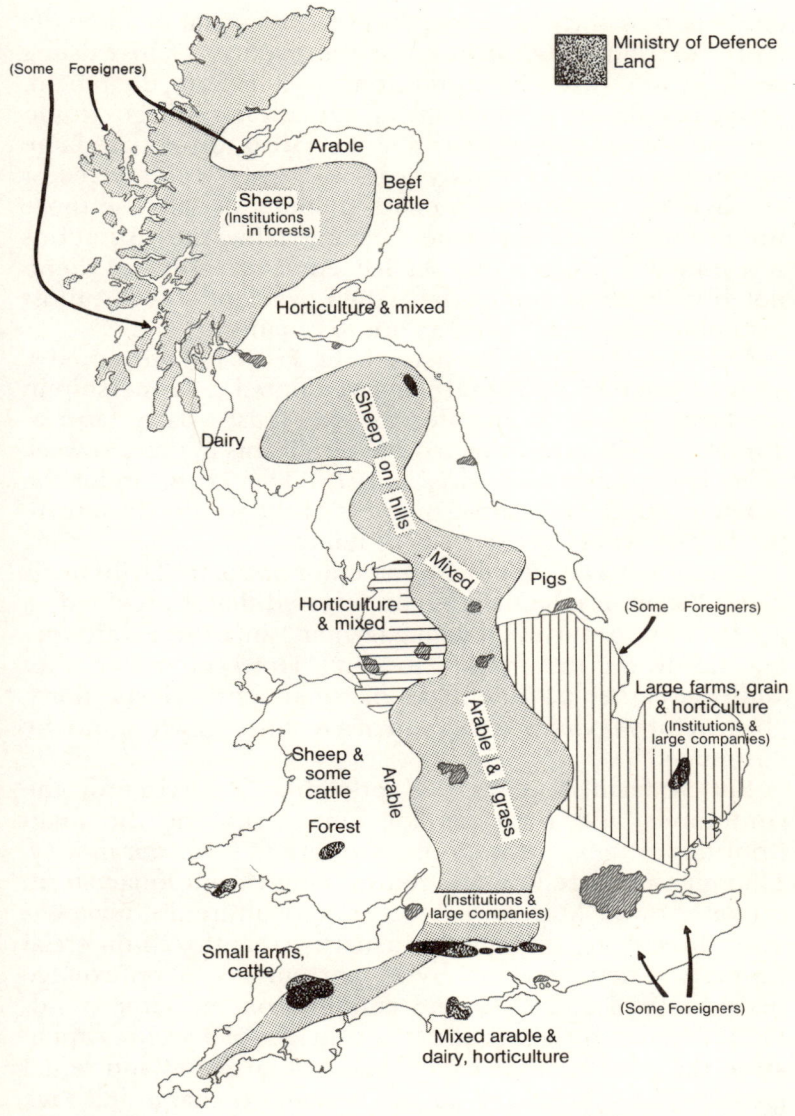

Ministry of Defence Land

(Some Foreigners)

Arable

Beef cattle

Sheep (Institutions in forests)

Horticulture & mixed

Dairy

Sheep on hills

Mixed

Horticulture & mixed

Pigs

(Some Foreigners)

Large farms, grain & horticulture (Institutions & large companies)

Sheep & some cattle

Arable

Arable & grass

Forest

(Institutions & large companies)

Small farms, cattle

(Some Foreigners)

Mixed arable & dairy, horticulture

12. Predominant Types of Agriculture: *institutional ownership is concentrated on the best land, notably in the East (see fig. 10). Increasing specialisation has pushed arable and grain-fed animals to the East and livestock to the West. (Source: M.A.F.F.).*

deal whereby the company provides the capital to allow the farmer to continue occupying, but losing control of, his existing land. In any case, the institutions say, without their help, young people and graduates from agricultural colleges would have no hope of entering farming. The break-up of many large private estates, thirty or sixty years ago, when prices were low and agriculture depressed, enabled young people and those without much capital to enter farming. Now, the institutions say, they will provide that service. Although some of them, notably Gallahers, want to farm their own land directly, most institutions let their holdings out to tenants.

The Pension Fund Property Unit Trust has proposed a scheme whereby young farmers are offered a partnership in the farm. Just such an offer for the Trust's dairy farm at Shrewton, Wiltshire, attracted 200 applicants in the first week of its being advertised early in 1978. The Trust, under the scheme, would keep three-quarters of the profits. In return, the farmer would get security of tenure.

Announcing an offer for a tenancy for one of the farms on its Guy's Estate, Prudential's agents stressed that it provided "a golden opportunity for a young farmer" since there were very few farms on the market to rent. However, to another audience earlier, a Prudential spokesman warned institutions: "be careful to avoid the creation of a tenancy which cannot be terminated".

But as farmers point out, providing a service for the community and rural employment cannot be the main priorities of these institutions. They are in it for the money. Though yields are low, the return comes from longer-term capital growth. Moreover, rents in agriculture are reviewed every three years, as against the usual five years for commercial property. They increased by 20 percent a year on average between 1976 and 1979 and continue on the same trend. PFPUT makes no secret of the advantages of a secure capital asset, the growth in the capital value of farmland and rental growth. The institutions do not hesitate to charge full rack rents. Warwick Thompson, farm fund manager of Property Growth Assurance argues, typically, that "a good tenant can afford a bigger rent and so protect an investment".

There are wider attractions in buying farmland. The Velcourt company acknowledged in evidence to the Northfield inquiry: "good quality land will be in strong demand, parti-cularly from institutional buyers. The price will bear much more relationship to returns than in the past and decisions will tend to be commercial rather than emotional". The British Insurance Association said that "growing fears of economic depression (makes it seem) prudent to invest modestly in a basic essential industry like food, which would still have a market if all else failed". The estate agents, Bernard Thorpe and Partners, says that "agriculture is a vitally important industry having the benefit of many grants and tax concessions not readily available elsewhere". Patrick Ravenhill, general manager of the British Airways pension scheme says: "we expect food prices to rise more than profits from manufactur-ing companies. Land," he adds, "does not depreciate in value like buildings". The institutions also point to good industrial relations in agriculture — a reference to the notoriously weak bargaining power of the National Union of Agricultural and Allied Workers, now submerged in the Transport and General, Britain's biggest union, with its widely scattered membership.

"Land", Rosen pointed out, "has consistently proved to be one of the few investments capable of maintaining its value even in times of high inflation". Yet his own Fountain Farming by 1980 had disposed of most of its acres with the company victim, it appeared, of the dangers of over-capitalisation and demanding or expecting quick returns on capital investment in agriculture. Eight of the company's farms in Lincolnshire and Hertfordshire were taken over by Booker Agriculture, part of the agribusiness company with widespread inter-national interests, especially in developing countries, and in the growing health food industry.

Foreigners

"We were told by one land agent that some overseas clients simply wanted to own a 'castle' and a substantial area of land; and that to them the type of land was often immaterial and might well be heather, rock and bog." – *The Northfield committee of inquiry into the acquisition and occupancy of agricultural land, July 1979.*

The control and ownership of land by large institutions and companies is widely acknowledged in both Europe and America to threaten the social fabric of whole areas and communities. So the ownership of land by foreign interests threatens the link, already tenuous, between the people and their nation. Yet Britain, and Scotland in particular, is at the mercy of any individual or corporation that can afford it.

The Northfield inquiry estimated in 1979 that at least 250,000 acres of British farmland had been bought by foreigners over the past few years, the bulk consisting of sporting estates in the Highlands of Scotland, and that overseas interests owned a total of up to 750,000 acres: 1 percent of the overall agricultural area but double that proportion in Scotland. In contrast to the situation in most other industrialised countries, farmland in Britain is at the mercy of 'the free market'. There are no restrictions on the purchase of British land by foreigners, whether they be tax-evading, offshore-based speculators, wealthy investors or simply working farmers who have run out of land in their own country and are attracted by relative cheapness of farmland in Britain. Questioned about recent purchases of farmland, Gavin Strang, a Labour spokesman on agriculture, told the Commons in January 1979 that "no information is available on whether the purchases were from within or outside the UK". The Inland Revenue admits that it is not possible to discover who is behind transactions if land is bought through nominees acting on behalf of foreign interests.

However, estate agents, the National Farmers' Union and the Royal Institution of Chartered Surveyors all agree that interest in British farmland by overseas companies and farmers is increasing. The estate agents, Strutt and Parker, said in 1979 that 10 percent of farmland coming onto the market was being bought by foreigners. A survey conducted by the NFU in England and Wales showed that in 1978 at least a hundred farms were owned by foreign interests, many of them in Kent, Sussex and Lincolnshire. Dutchmen in particular have concentrated on Lincolnshire where some of their countrymen settled in the middle of the seventeenth century. (It was the Dutch, too, who pioneered techniques which enabled the fens to be drained and who built up the East

Anglian bulb industry.) According to the NFU, Dutchmen also own forty-eight farms in other counties. Derek Turnbull, a Norwich estate agent, noted in 1979 that the Dutch were buying up large farms in the area, and pointed out that with Norwich airport providing daily flights to Amsterdam, many Dutch families were settling in East Anglia and commuting to Holland.

Though the Dutch have been in the vanguard, Germans, Danes and inevitably Arab interests (for the most part interested in residential property in the country) have steadily increased their stake in British farmland. To take one example: in the summer of 1979, a 1,000 acre estate, including a 200-year old mansion, Shotesham Park, near Norwich, was sold to a group of Danes for £1.6 millions.

Some foreign buyers are sensitive to local needs and feelings, some are not. In England, half the population of Woodmancott, a village near Basingstoke, was evicted in 1976 after the sale of the farm where they worked. Seven tenants were forced to leave. The farm and its eight cottages were sold by Winchester College to a Danish farmer, Anders Dineson. He brought in Danish labour and, according to the local vicar, the village was destroyed as a community.

What evidence there is suggests that Britain will be treated increasingly as the haven for foreign investors and for farmers facing tighter restrictions in their own countries and elsewhere. One of Britain's attractions is the relatively low cost of land, a third or even a quarter of the price on the Continent. With superb irony, given the seizure of Irish land by successive British administrations from the Norman to the Cromwellian and later farmers in Ireland — where the impact of the EEC has not only brought unprecedented wealth to the agricultural sector but also a dramatic, if temporary, rise in land prices — are buying up farmland in Britain. Farmers' Business Development, Ireland's largest agribusiness group owned by farmers and chaired by Paddy O'Keeffe, has bought 1,000 acres in Oxfordshire. The Ministry of Agriculture in Dublin points to the example of one Irish farmer who in 1978 bought three times as much land in south-west Scotland with the money he got by selling his farm in County Meath. A Dutch farmer told the Northfield committee that he had been able to buy a farm in England three times the size of a similar one in the Netherlands for the same amount of money.

Foreigners sometimes buy land in Britain to avoid tax. Shareholders residing overseas are not subject to Capital Transfer Tax, and do not have to pay Capital Gains Tax when disposing of the land. Companies buy land and subsequently sell it at significantly higher prices to associated companies.

Estate agents, who have a strong vested interest in promoting the free market philosophy, argue that any attempt by Britain to introduce limits on farmland ownership by overseas interests would provoke retaliation against British investors in the commercial property market in Paris, Amsterdam or Brussels (where as speculators they have been more active than successful). The agents say that EEC rules covering freedom of establishment and free competition prevent Britain from introducing any restrictions against people from other Common Market countries. This is nonsense. Indeed, defenders of 'a free market' acknowledge that one of the real problems in the EEC is that the same rules must apply equally to foreigners as to British subjects. Thus the trouble about imposing a residence qualification is that it would limit the activities of British absentee landlords. And any limits on the amount of land any foreigner could own would have to apply to British landlords. Yet even these assumptions do not stand up to a study of the practice in other Common Market countries.

In the United States, where the principle of the open market is adhered to more than in most other countries, concern about the amount of farmland being bought by foreign interests led to the 1978 Agricultural Foreign Investment Disclosure Act requiring foreign owners to register their holdings with the government. The Act also set up a system to monitor the effect of overseas purchases of land on family farms and rural communities. In both Norway and Finland permission by the government is needed before a foreigner can purchase farmland. New Zealand and Switzerland demand residence qualifications.

In most EEC countries legislation is designed to control the activities of any large corporate group, whether foreign or home-based. But controls, some of which are highly sophisticated do vary.

In Denmark, there is an upper limit of 185 acres on the amalgamation of holdings; no individual can own more than two farms, and the second farm can be no more than 15 kilometres from where the owner lives. Farming must be the main occupation of the landowner. People wanting to buy land must satisfy the State Land Commission that they have farming experience which must be both adequate and equivalent to that obtainable in Denmark — a pretty effective barrier against foreigners.

In France, there is a network of controls aimed at both foreigners and French interests. Legislation, designed to promote family farms, prevents land coming onto the market being bought by a farmer whose holdings are already above a given maximum, while large blocks of land are split up if they exceed by three times what is regarded as the viable minimum for a farm in that area. In the early 1960s the French government set up a series of locally-based Land Improvement and Rural Settlement Companies — known as SAFERS under their French acronym — primarily to buy up land and sell and distribute it to help farmers establish themselves. These organisations have a pre-emptive right to purchase land coming on the market and to hold it for five years. They can help young farmers as well as those whose land has been used for urban development. Government departments can consult local SAFERS when a foreigner applies for permission to bring in capital to buy a farm. The company can advise the government to refuse permission in spite of EEC rules on the free movement of capital. France retains the right to impose exchange controls. The Northfield committee visited the SAFERS at Blois in the Lois and Cher department and was told that it had excluded, for instance, a group of Japanese who had wanted to buy a vineyard in the region. Northfield added: "because its representatives were informed of all land transactions which took place, they could make it very difficult for a foreigner to purchase agricultural land; but (the SAFER) added that if a substantial block of land came up for sale, they would almost certainly intervene, regardless of the nationality of the purchaser".

In West Germany, there is a strong cooperative tradition among farmers. The Land Settlement Society there may veto

the sale of land on the grounds that the price is too high. It
buys land and allocates it to farmers in needy regions, and has
the authority to prevent large farmers from increasing their
holdings.

In the Netherlands, the tenant has the first choice of land
sales and there is growing pressure for strict residence
qualifications. The government has set up a national body to
which all farmland sales and transactions will have to be
referred.

In Ireland, the Land Commission buys and redistributes
farmholdings and its consent is required for any significant
sale of land. When Ireland joined the EEC, the government
agreed to relax its controls on farmland ownership by
foreigners by opening the market to them in land that had
been abandoned or left uncultivated for more than two years
or where the foreigner had worked or provided work in the
country for two years. It was not liberal enough for the
European Commission. But there are many ways, as the
French above all have demonstrated, of paying respects to the
principles of the Common Market law while avoiding them in
practice. In any event, on such a fundamental issue as
landownership, there is absolutely no question of EEC
institutions or other member governments preventing another
government from adopting its own legislation on land,
particularly if there is strong popular support for such a move.
Landownership is much more of a controversial, emotional,
and, indeed, serious, question in other countries than in
Britain with the possible exception of Scotland. Yet there is no
reason why this should remain so: it is time that elected
authorities and local communities here adopt a more hard-
headed look at the prevailing British attitude and the complacent 'gentlemanly' approach of Whitehall towards farmland
ownership and the aggressive support of the 'free market'
philosophy of such vested interests as estate agents.

3

THE MARGINAL AND FORGOTTEN AREAS

"The view has been expressed that the Highland area is the only unspoilt region in Europe and should be kept in its present state. This implies that the presence of man and agriculture is in some way undesirable, and this I would question." – *Dr John Cunningham, director of the Hill Farming Research Organisation, Midlothian.*

Scotland

The history of landownership in Scotland is particularly marked by greed, brutality, speculation and neglect. It has witnessed the Highland Clearances, depopulation, absentee landlords, tourist and forestry development, the turning over of cropland to heather and game shooting, the entry of foreigners to buy up estates previously owned by individuals who for financial reasons, or simply lack of interest, allowed the land to wither.

Landownership is a more controversial issue in Scotland than in the rest of Britain, a vestige of the misuse and abuse of the land, including the Clearances, and a consequence of a relatively small number of large estates, some of which are used for the sole purpose of sport. Landownership has also been made a public issue both by the Scottish Nationalist Party and by the Church of Scotland. The complacency and lack of knowledge about landownership in the Scottish Office and its lack of concern reflects the prevailing approach in the rest of Whitehall. Such an attitude means either that the status quo is preserved, or that change is determined by the only groups which can act: foreigners, large financial institutions, and investment trusts.

The growing number of people who are trying to question this state of affairs, and seek information, have been helped by a brilliant piece of detective work and a quarter of a century of painstaking research into old records, land registers, surveys and maps by a 90 year-old former estate worker and forester. John McEwen could conclude that 100 landlords own a quarter of Scotland, nearly five million acres.[1] At the time of his research these were the leaders:

Duke of Buccleuch	277,000 acres
The Wills Family	263,000 acres
Lord Seafield	185,000 acres
Countess of Sutherland	158,000 acres
Duke of Atholl (the nephew of Lord Cowdray)	130,000 acres
Captain Alwyne Farquharson	119,000 acres
Duke of Westminster	113,000 acres
British Aluminium	110,000 acres
The Earl of Stair	110,000 acres
Sir Donald Cameron	98,000 acres
Duke of Roxburgh	95,000 acres
Lord Vestey	93,000 acres
South Uist Estates Limited	92,000 acres
Lord Cowdray	88,000 acres

Other large landowners include: the Duke of Argyll, with 74,000 acres, Lord Home, with 54,000 acres, the Duke of Portland, with 48,000 acres, Lord Thorneycroft, chairman of the Conservative Party, with 44,000 acres, Lord Cawdor, with 39,000 acres, the Eagle Star Insurance Company, with 30,000 acres, and Lord Leverhulme, with 23,000 acres. The Marquis of Bute still owns the island of Bute (30,000 acres). The Wills family, headed by Lord Dulverton uses most of the land for sport and forestry, while Colonel Whitbread, of the brewing family, is the largest landowner in Ross and Cromarty. It is there, McEwen notes, "that the secretive Arabs are now settling in". Whitbread himself in the late 1970s offered land to prospective buyers on the Continent and in the Middle East.

The Queen's Scottish estates, around Balmoral, are larger, by the way, than those of Queen Victoria, her great-great-grandmother, who was particularly attracted to Scotland.

340 families and companies own two thirds of the Highlands and Islands — Caithness, Sutherland, Ross and Cromarty, Inverness, Argyll and Perthshire. 0.1 percent of the population there owns 64 percent of the land.

We have noted the Northfield committee's estimate that foreigners own 320,000 acres in Scotland and much of this is in the Highlands. The Chief Valuer for Scotland acknowledges that there were at least 54 purchases by overseas buyers in Scotland between 1970 and September 1978, accounting for 230,000 acres. Most of these were in the Highlands and Islands and most consisted of sporting estates. The committee said that this was likely to be an underestimate since "some overseas buyers will have been overlooked through being incapable of identification as such and some purchases will have taken place prior to 1970". The Scottish banks said they were aware of 66 purchases by foreigners: 30 by Dutchmen, 12 by Americans, 12 by Arab interests, and the remainder from a number of other countries. The National Farmers' Union of Scotland spoke of an 'accelerated increase' in purchases by foreigners in 1978.

The Perth-based estate agents, Bell-Ingram, who actively go out and cultivate foreign clients, had a current list of interested buyers in 1979 that included 27 Dutchmen, 5 Belgians, 5 Danes, and 5 Germans. In the course of an investigation it conducted in 1978, *The Scotsman* estimated that over the previous decade at least half a million acres in the North of the country and in the Islands had been bought by foreigners, and that up to 1 million acres were now owned by interests based outside Scotland. Between 1975 and 1977, 20 Highland estates were bought by overseas interests, mainly Dutch, German and Swedish. Nearly twenty years previously the Swiss Panchaud brothers bought the Mar Estate, near Braemar, and turned it into a sporting estate. They also let for sporting the Tulchan Estate, previously owned by Jim Slater.

But the Dutch have been in the vanguard, buying possibly as much as 200,000 acres over the past five years. Some of them use the Highlands as the base for trade in meat, especially veal, while other, equally enthusiastic entrepreneurs, have invested in deer forests, chalets and holiday attractions. Some have quickly sold off parts of estates at a large profit. Some have proved to be good and enterprising farmers — growing vegetables on land which lay fallow before. Others have shown little respect or understanding of the rights of crofters. Johannes Hellinga, a Dutch millionaire and developer, is one of the most controversial.

Since 1976, for just over £1 million in all, Hellinga has bought 16,000 acres. Of this he sold one third immediately, upsetting and disrupting local communities in the process. Five months after buying Kindeace House and 7,000 acres near Invergordon — where he quickly gave notice to sitting tenants — he sold 4,000 of these acres to the Forestry Commission. But Hellinga, who also bought the 9,000 acre Waternish Estate on Skye, has succeeded in growing vegetables there for the first time. He offered much of this estate in parcels to crofters for 5p an acre, an offer they rejected mainly, it appears, because they resented his plan to group them together into a kind of management cooperative.

Another Dutch businessman, Sijtse Kats, has purchased an estimated 40,000 acres including the Belladrum Estate, part of which he sold back to Scottish farmers; Durness, where he wants to develop fluorspar mining; Torridon, where he is fighting the Countryside Commission over his plan to develop a holiday camp; and Rogart in Sutherland.

Teakele Soetboer, a Dutchman who owns a chain of hotels in the Netherlands, bought the Brin Estate near Inverness for £57,000 in 1977, and he, too, planned to build a range of holiday chalets there.

Like Hellinga, Keith Schellenberg, who became the laird of the Isle of Eigg which he bought for £265,000, has developed the agricultural economy of the island. But if the islanders do not like his plans, then they have no alternative but to leave.

Estates bought by Dutchmen also include the Isle of Erraid, the Strome deer forest, Bendamph, Fannich deer forest, Foich Lodge, Braemore, Forsinard, the Tressady Estate, Edderton, Coignafearn and Blervie.

Despite the possibility of investigation, albeit in a piecemeal way, the Scottish Office prefers to rest on its ignorance. Asked how much land foreigners have bought in Scotland over the past few years, it replied: "We are not able to comment . . . nor can we indicate how much land involved is actually agricultural land". "Is the Government concerned about the change in ownership?", I then asked. "Foreigners who buy land are subject to the same constraints as applied to any other owners of land in Scotland", came the reply.

The Dutchmen's success, though questionable, is also an indictment of the way Scottish landowners have treated the land in the past. The Department of Agriculture in Scotland pays out more than £50 millions a year in subsidies and grants, yet landlords continue to opt for the easier course of letting out their property for sporting, with some of them going to the extent even of turning cultivated land to heather to encourage grouse breeding.

On the Vestey estate in Sutherland, McEwen says: "Most of the (Stoer section) is held for sport and so the less developed and the less populated it is the better. It suits absentee landlordism admirably". In Perthshire, the Stuart Fotheringhams of Grandtully and Murthly, he notes, let part of their 45,000 acre estate to an American syndicate as a game reserve. Of the Meggernie Estate in Glen Lyon, Perthshire, he reports: "agriculture and forestry sacrificed to game . . . it is sad that the game racket is responsible for the degradation of very extensive areas of our valuable grazing and forestry lands, some of which are the richest in Scotland".

The Highlands and Islands Development Board (HIDB), which has been persistently sabotaged in its ambitions by the lairds and the Scottish Office, have told successive Governments about how tenant farmers are forbidden to improve marginal hill land because of the threat to the "nesting habitats of grouse", of others being forced to accept short leases so that owners could "capitalise on the market whenever it seems most profitable". Describing incidents on the Island of Mull, Sir Kenneth Alexander, the HIDB chairman, spoke to the Scottish Landowners' Federation in 1977 about the "acute under-use of land resulting in loss of jobs to the extent that a whole community is undermined and crippled".

During the Highlands clearances, more than 1,000 families were evicted by the lairds of Mull to make way for sheep. Now the island is suffering from a more subtle form of decline. Instead of being evicted, the shepherds and farmworkers are given notice to quit. They are being cleared to make way for the pursuit of hunting, shooting and fishing, and for holiday-makers. The result is homelessness. Over 15 years ago the Government was warned in a report that "unless something radical is done, the Island will have declined to a point at which a revival of community life, and a viable economy, might no longer be possible". Agricultural decline continues. The future of Killiechronan — one of the Island's most productive estates — and the 40 people who live on it were threatened a few years ago when the former owner, Mr Henry Hood-Barrs, died in his eighties. The HIDB, aware of the successful land reclamation schemes that had already been implemented there, offered to provide £275,000 to help. The Government vetoed the idea. In came Mr David Holman, a London insurance broker. He paid an undisclosed sum for the estate (with the exception of 3,000 acres sold to the Forestry Commission), told the shepherds to sell Killie-chronan's 3,500 ewes, made them redundant, and put their cottages on the market for a price of up to £18,000 each.

"My whole policy", Holman said, "has been to concentrate in one area, or two, instead of trying to look after everything and lose money on it all. I admit it is probably different to what the Government would do with your, or my money". Yet this estate once boasted 6,000 sheep, 400 cattle and several hundred acres of arable land. John MacRae, a shepherd for twelve years, managed to rent a holiday home but had to get out before the beginning of the summer. His son, Jimmy — a carpenter — found a job at the neighbouring 14,000 acre Torloisk Estate, owned by Captain Alwyne Farquharson, sixth biggest Scottish landlord on McEwen's list, but then he was told the estate could not afford the luxury of a joiner.

Killiechronan and Torloisk are two of the six large agri-cultural and deer-stalking estates on Mull. In all they cover about forty percent of the inner Hebridean island. Another 20,000 acres are owned by about 10 landowners, and much of the rest is controlled by the Forestry Commission. Other islands — Pabay, Staffa, with Fingal's Cave — are offered for a few hundred thousand pounds or less. "The unacceptable face of feudalism," says Sir Kenneth Alexander, "is, when it appears, if anything even less attractive than that other physiognomy to which Mr Heath referred".

The most modern industries — nuclear, with the fast breeder prototype nuclear reactor at Dounreay on the Caithness coast, and the prospects of pits for nuclear waste, and oil — have come to the Highlands and Islands. But it was succinctly put by the Fraser of Allander Institute for Research, in a study on the Scottish economy. It described the Highlands as an exploited area with a 'boom and bust' economy supplying raw materials and surplus labour to industrial centres when required. Export earnings, it said, of cattle, kelp, sheep, herring and oil are not reinvested in terms of productive assets or alternative industry. Shetland may be getting richer, but it is suffering from the scars of development. As one correspondent put it, fishermen are working as well-paid menials, crofters are giving up a traditional pattern of life to work long hours with the oil industry. The young love it — for the moment.

A perfect but depressing example of how investment decisions are made and broken came on 30 December 1981 when the British Aluminium Company suddenly announced the closure of its smelter plant at Invergordon on Cromarty Firth. The decision meant the immediate loss of 890 jobs; the only other significant employer in the town is the whisky distillery.

The smelter was a symbol of the new industrial future for the Highlands. Mrs Isobel Rhind, a local councillor, recalled a meeting she held in her home fifteen years ago: "The whole area had become apathetic, I could see little hope of future employment with farms becoming bigger and more mechanised, and the corresponding loss of jobs. I could see that with a combination of deep water, flat land and electricity it was an ideal place for a major industrial base".[2]

The decision has much wider implications, especially in view of the Government's determination to press ahead with nuclear power and the nuclear industry's insistence that their power will be much cheaper than conventional power. The smelter was to have been provided with 'cheap power' from the Hunterston B nuclear station in Ayrshire. Persistent problems with the station meant that the company had to buy expensive power from the Central Electricity Generating Board in England.

Loch Kishorn caravans. *Caravan accommodation for workers temporarily employed on the construction of an oil platform by Loch Kishorn in North West Scotland. But they have no roots there and their employers have no plans to provide long-term employment in the area. (Photo: Don McPhee)*

In the meantime, the remaining 18,000 or so crofters dispersed among two million acres who rely on Government support are under attack. Westminster and Whitehall resent the cost and the Thatcher administration was pressing to sell many of the Scottish Department of Agriculture's 2,800 holdings. And while owner-occupation of crofts is growing, more and more land — 8,500 acres in 1979 — is being taken out of crofting, to forestry, holiday chalets and caravan sites.

Against this background of speculators' asset-stripping, of land abuse and mismanagement and depopulation, the HIDB, whose limited loans have already increased employment in some areas, is now pressing for some additional powers to enable it to intervene directly in the land market, in disputes over land use, for example, and particularly those involving absentee and foreign landlords. It proposed designating areas of about 50,000 acres where landowners would be offered financial assistance to improve the land and offer, in turn, leases to young tenant farmers. Reserve powers of compulsory purchase would be used only in the last resort. These powers already exist in legislation though they have never been used. Even though the HIDB considers itself to be essentially a development board, promoting employment and better land use, rather than a statutory landowner in its own right, opposition to it by large private landowners has verged on the hysterical. They are particularly concerned about the notional threat to their freedom against the background of the growing interest in their estates by wealthy foreign and institutional interests. Their estates may be run-down, but they are becoming hot property.

"There are some who regard the Highland region as little more than a wilderness and consider that agriculture, especially hill farming, is of no consequence" (and should be substantially abandoned), John Cunningham told the British Association in 1974. "We may be discussing an underdeveloped area, but certainly not one in which the spectre of starvation haunts the inhabitants, but rather one which could make an increasing contribution to the national larder for the teeming millions packed in the south-east of England".

Governments of both main parties have said: "Public opinion is not yet ready for Government interference in how

land is used in Scotland". That, of course, is an excuse against
more public and democratic participation in decisions about
the fundamental issue as land use. And there can be no
argument against urban Scots, around Glasgow, for instance,
ancestors of those who lived on the land, having the right to
live a better life and claim the open spaces to the north.

Wales

Wales, as its geography and landscape suggests, with its small
fields and poor land, is largely a country of hill farmers and
smallholdings. To the traveller entering the country from the
wide plains and proud manor houses of Herefordshire the
difference is striking. Three-quarters of the land in Wales is
owner-occupied, and over 20 percent of the country consists
of 'marginal land'; that is, land that could be used, given
adequate resources and a change in the priorities of those in
Whitehall and Brussels, for more productive agriculture.

The recent history of Wales is a classic example of economic
and political abuse: the break-up of traditional and local
communities as a result of the capitalists' search for quick
profits and wealth. As soon as coal was discovered under the
South Wales valleys, the landowners enticed the predominantly
rural population to work underground. The rows of cottages
scattered along the hillsides and high unemployment
now bears witness to the way people have been treated, as
temporary wealth-creating instruments.

The countryside is left behind them deprived and depop-
ulated, an attraction for tourists and the Forestry Commission.
An increasing number of Welsh sell off their farmhouses and
cottages to people from English towns. North-west Wales has
the dubious distinction of providing the highest density of
second-homes in Britain. In the middle of Anglesey, its chalets
and holiday homes, lie 16,000 acres of rough farmland, lying
in wait for a property speculator. In Wales, the number of
second homes matches the number of homeless.

Wales does not belong to the Welsh, any more than
Scotland belongs to the Scots. That applies to the land, the
resources under the ground and even the air. And as we have
noted virtually the whole of Welsh airspace is controlled by the

Ministry of Defence which uses it for training.

There may be an alternative future for Wales. But a report published by the University of Wales Press in 1978 — "Attitudes and Second Homes in Rural Wales" — suggests that people have learnt to accept the influx of the urban English for the most part with a passive shrug of the shoulders. And other interests are moving in. 13 percent of the land area is already owned by large financial institutions; estate agents are encouraging overseas interests, including Arab-based, to buy property. Well over half the number of farmers are more than 55 years old. Hill farms come up regularly for sale.

The Hills

> "It was remarked in 1954 that in parts of the Scottish Highlands where there were hedges and fuschia and where palm and eucalyptus grew in the open, it was all but impossible to obtain locally-grown fresh fruit and vegetables in the month of August, with bread bought from Glasgow and fish from Aberdeen" – *E.J.T. Collins, "The Economy of Upland Britain", Centre for Agricultural Strategy, May 1978.*

The hill areas, which once played a key role in the country's economy, fell victim about 100 years ago to the forces of free trade, encouraged by Britain's imperial ambitions, and to the industrial revolution. They were treated as a source of manpower for expanding cities; one of their chief attractions now for townspeople is their deserted landscapes and 'unspoilt countryside'. They have become dependent on hand-outs and subsidies from central government, just one of many calls on public expenditure. Britain is in a unique position in Europe in enjoying a hill farming economy providing breeding cattle and sheep for consumers in the lowlands and in the industrial heartlands. Yet this traditional role is also under threat, neglected, as British agriculture becomes increasingly dominated by the interests of capital-intensive farming. Pricing policies and agricultural research are geared to meet the needs of intensive livestock rearing and cereal growers.

In the early stages of the Industrial Revolution, the uplands of Britain provided cheap water power from the fast-moving streams for the textile mills. As late as 1850 more than half the world's production of zinc, lead, and tin were extracted from British ores, mainly in the hill areas. In 1910, the Scottish Highlands produced 20 percent of the world's aluminium. Local industry, for a brief period, stimulated local agriculture. Then the economy of the hills collapsed in the face of competition from new manufacturing centres based on coal and the railways in the lowlands and on the coast. The policy of free trade, promoted in Britain at a time when other European countries were protecting their economies, gave cheap raw materials for industry from overseas and cheap food for the rapidly growing urban population, but compounded the problem. The population of the hills, tempted by the relatively high wages industry could offer, drifted to the mass consumer markets. By 1950, less than five percent of the total population lived in the uplands compared to 20 per cent two centuries earlier. Less than 20 percent of the population of Wales lived in the industrialised counties of Glamorgan and Monmouth in 1800; by 1950, these two counties accounted for a full 60 percent. In some hill areas of Britain, the population fell by 25 percent over the past ten years; the trend continues.

Another example signals what has happened, and is still happening, to these regions of the country: in 1901, less than 2 percent of the male labour force in Merioneth, mid-Wales, was employed in public administration and defence. Fifty years later, that proportion increased to more than a quarter, and accounted for more than agriculture and forestry combined.

The problem now is the same as it was a hundred years ago. As Collins quoted above, points out, it is to define "the local as against the national interest, to reconcile them, and to decide as a matter of policy whether to reverse the historical trend of upland migration, or whether to allow the level of economic activity and of population to be determined by market forces".

As sheep took over from people, so deer and forestry are taking over from sheep. The fast-growing tourist and leisure industries have encouraged hill farmers to offer accommodation and other attractions to temporary visitors, with the danger that farmers are finding they can get more money from the tourist than from agriculture. Yet according to John Cunningham of the Hill Farming Research Organisation, over a million acres of lowland Britain would be needed to replace even the present agricultural output of the uplands. Hill areas, for instance, produce about half of Britain's lamb. Local fishing, too, has suffered in face of the armies of sportsmen. Fish farms have been set up by large corporations such as Unilever, with an eye on the demands of expensive restaurants. Large trawler owners based in Aberdeen and Humberside, pushed out of grounds further afield, off Iceland and in the Arctic, are threatening inshore fishing fleets on which small, local, communities depend. The EEC is just another threat.

The 16 million or so acres of hill land — inhabited by just 50,000 families — could produce much more food. John McEwen estimates that the productivity of Scotland's hills, for example, barely reaches fifty percent of its potential, in some areas, and that estimate is shared by many agricultural colleges. Agricultural research, however, has concentrated on producing hybrid, high-yielding cereal varieties suitable for lowland arable land; drainage and irrigation of uplands, and grassland production have been consistently ignored even though grass relies merely on rain and sun, not chemicals. Of the £11 millions in aid granted in direct subsidies to British agriculture by the EEC in the late 1970s, Wales got less than £150,000.

The neglect of communities and the productive and natural wealth of hill areas is another symptom of the division between town and country, of the material and bureaucratic pressures on the population to congregate in large cities and accept without question the demands of an increasingly monopolistic industrial society that seeks the maximum and quickest profit from the land.

There is one other pressure on hill areas that is certain to increase: the forestry lobby which seizes enthusiastically on concern about Britain's future timber needs.

Britain imports about 92 percent of the wood and wood products it consumes at a cost of nearly £3,000 millions a year. It relies on wood imports significantly more than any other EEC country with the exception of Ireland and The Nether-

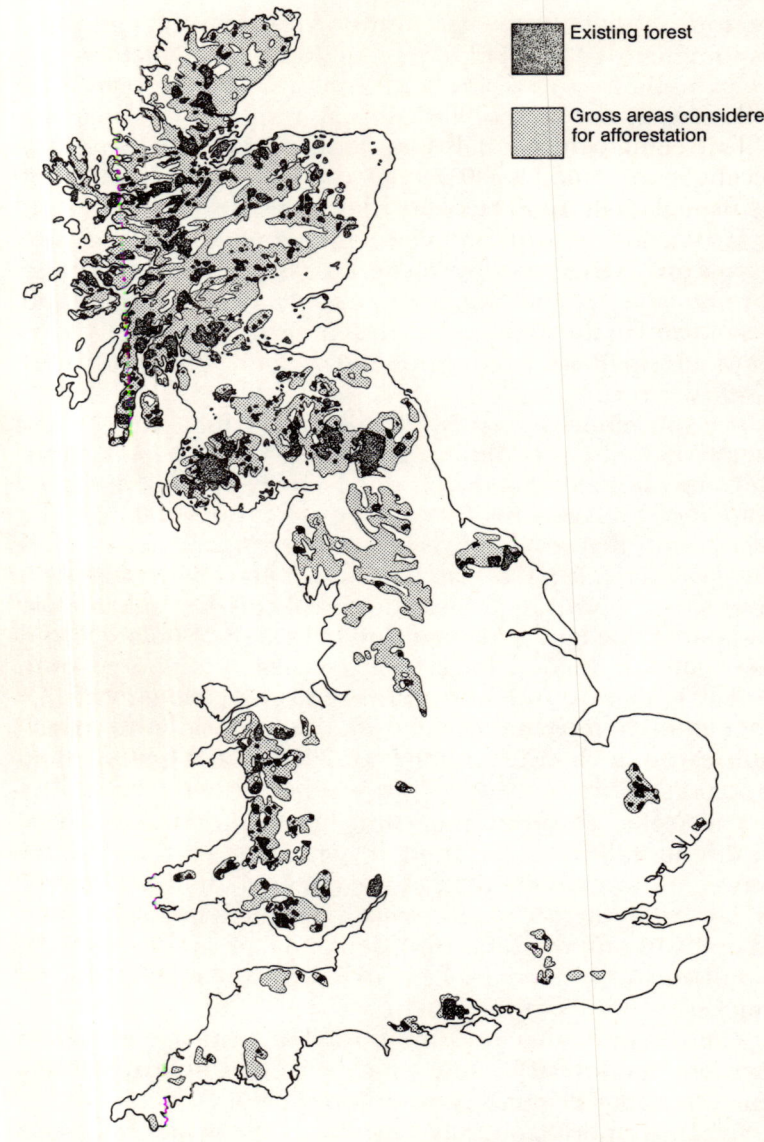

Existing forest

Gross areas considered for afforestation

13. **Afforestation:** *an increasing amount of land will be taken up by forestry, which threatens farming. One solution to the controversy is to mix livestock and trees.* (Source: Forestry Comm.)

lands. It would, therefore, be particularly vulnerable to world shortages widely forecast by the turn of the century. Prices of timber are assumed to rise by 30 percent (in excess of inflation) by the year 2000.

Just eight percent of Britain is forested, even though in recent years some 75,000 acres annually have been taken up by new plantations. The controversy over forestry — centred in particular around the conflict between sheepfarmers, notably in Wales, and the Foresty Commission and between the Commission and conservation groups — took a significant new turn in the beginning of 1980 when the Centre for Agricultural Strategy (CAS) proposed an ambitious plan to double the present five million acres of plantation (half of which is owned by the Forestry Commission, half by private landowners) by the year 2030. This would mean the loss to farming of a third of all the hill areas — what the CAS describes as "unproductive moor" and mountainside. The CAS also estimates that there are about 500,000 acres of "unproductive and decaying woodland and scrub within farms in England and Wales" — the trouble is that many farmers are unaware of either the economic or environmental potential of woodland.

The CAS proposals were only slightly more ambitious than the Forestry Commission's own plan which includes the afforestation of a fifth of the 11.8 million acres of 'rough grazing' land in Scotland. The plans were bitterly opposed by conservation groups: the Council for the Protection of Rural England said that they posed a dangerous threat to National Parks and common land. It pointed out that the forestry lobby, which wants the Forestry Commission to have wider powers, had already argued that large areas of Dartmoor, the Yorkshire Dales, the Pennines and the Scottish uplands were 'wasted'.

One of the main objections is the planting of a monoculture of conifers — notably Sitka spruce — which are not indigenous to those areas and which would threaten wildlife, including such birds as the merlin, the peregrine, dunlin, curlew, golden plover and golden eagle. In addition, intensive plantation would mean a persistent use of machinery, fertilisers and pesticides. Water supplies could be threatened.

Poor Forestry Planting. *New forestry plantation near Devil's Bridge in Wales. Brutal changes in the landscape have provoked storms of protest from both sheep farmers and conservationists. (Photo: J.A. Taylor)*

Good Forestry Planting. *Forestry planting can blend sympathetically with the landscape. This is Guydyr Forest in Wales. (Photo: The Guardian)*

The Forestry Commission's crude policy of promoting blanket coverage of fast-growing conifers has given it many enemies. In the Forest of Dean, the Commission has fenced and sold off land — which for hundreds of years has been open to the public — to private investors. It has felled a large number of deciduous trees, including oaks, to make way for pines. With little real consultation with local interests, it has drawn a line along Welsh hills, changed the nature of the landscape, restricting grazing area for livestock, and not even contributed to rural employment and the local economy. I shall never forget the disdain with which a local Forestry Commission representative was greeted by Welsh hill farmers one evening in a pub in Tregaron.

Though investors in forestry benefit from generous tax privileges and grants — paid out of public funds — and though targets for afforestation are invariably conservation areas whether National Parks or Sites of Special Scientific Interest, forestry plans are not subject to normal planning controls. The Forestry Commission is obliged only to consult local interests and conservation groups.

New plans to afforest part of the Brecon Beacons National Park provide a classic illustration of the conflict between forestry and conservation interests. The proposals, to plant conifers on about 650 acres of the Cnewr estate near Sunnybridge, incensed the Countryside Commission and the Royal Society for the Protection of Birds, yet the plans were backed by the National Park committee. The Countryside Commission, argued that the plans would change the whole character of the mountains and in particular their high, open, spine. The attraction of the open spaces, it pointed out, was the main reason why the area was designated a National Park in the first place.

The controversy over forestry, like so many debates in Britain, has become so polarised, such a confrontation, that is is difficult to see how suspicion between the two sides can be overcome and make way for compromise. The forestry lobby quotes forecasts suggesting that demand in Britain for timber will increase by 50 percent by the year 2000 by which time, as we have noted, the world price may rise by one third. Yet other forecasts suggest that the economic situation will mean that

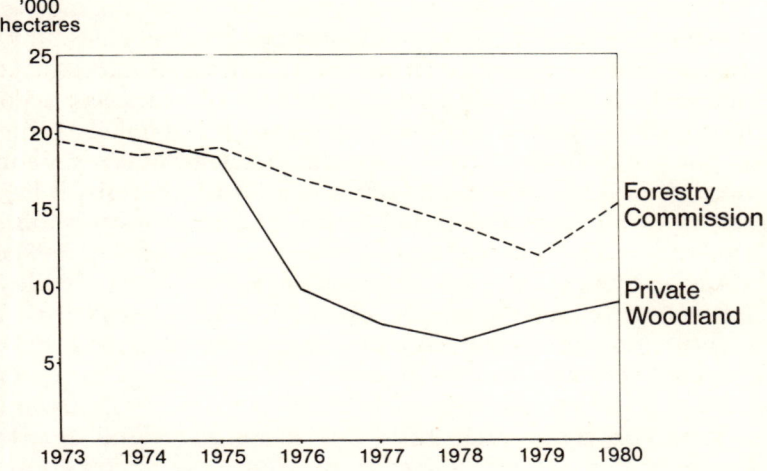

14. New Forest Planting.

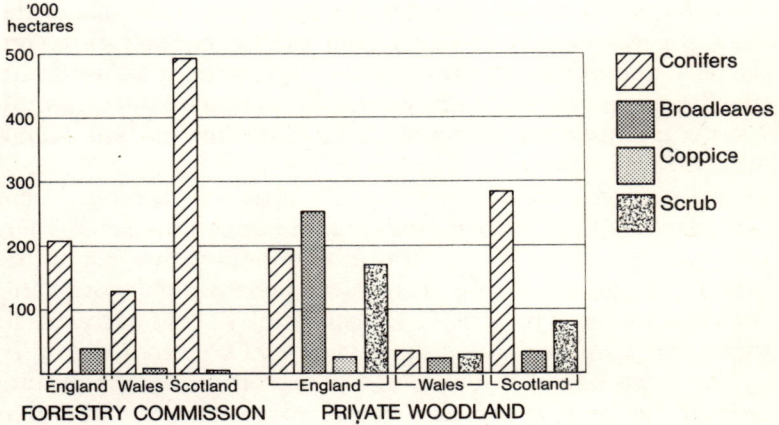

15. **Forest Areas under Public and Private Ownership:** *Britain imports more than 90% of the wood and wood products it consumes each year at a cost of some £3,000 million. Land devoted to forestry is lost to agriculture, often exchanging open moorland for uniform conifer plantations. (Source: Forestry Comm.)*

Britain's requirement for softwood for construction, for instance, will not increase at all after the first few years of the 21st century. The use of paper and newsprint could be reduced significantly as technological developments in communications are introduced.

The question of Britain's future forestry pattern and its needs for such a basic raw material as wood is so neglected yet so important that it is crying out for a public inquiry on a national level. There is one obvious solution, much talked about but where little has been achieved. That is, to establish less-intensive and relatively small woodland areas with a combination of conifers and traditional broad-leaved, deciduous, trees. Discouraged, perhaps, by the propaganda of the Forestry Commission, its threats of compulsory purchase orders, and claims that there is a 'shortage of land' readily available, both farmers and amenity groups have been slow to recognise the potential value of integrating farming and forestry, especially in hill areas.

New Zealand for many years has been developing ways to combine forestry plantations and sheep farming. It is not difficult; all it needs is a little care to protect the bark of the young trees. A survey of about 20 farms in Caithness conducted by Edinburgh University's forestry department showed that on holdings which had increased tree plantation by 30 percent, lamb production doubled. Tree shelter had been a contributory factor.

Britain could also learn the lessons of Sweden where farmers' cooperatives own both timber plantations and timber-using industries. That is a more labour-intensive, economic and politically acceptable response to large timber groups and investors here. The British forestry lobby seeks recourse to arguments about the 'national interest'. Of course, the argument exists but it is up to all interested parties, including elected respresentatives and delegates to establish where that interest lies. There is also a world-wide dimension to the problem (see chapter 11). The equally cynical attitude, of greed and short-term considerations and of giving freedom to 'market-forces' is allowing international timber and food companies to cut down in the world's tree population, including valuable tropical rain forests, at a rate of 50 acres a minute. Not only does this have far-reaching implications for soil, water and agriculture; tropical rain forests are the guardians of thousands of animal and plant species and a unique source of medical, including anti-cancer, drugs.

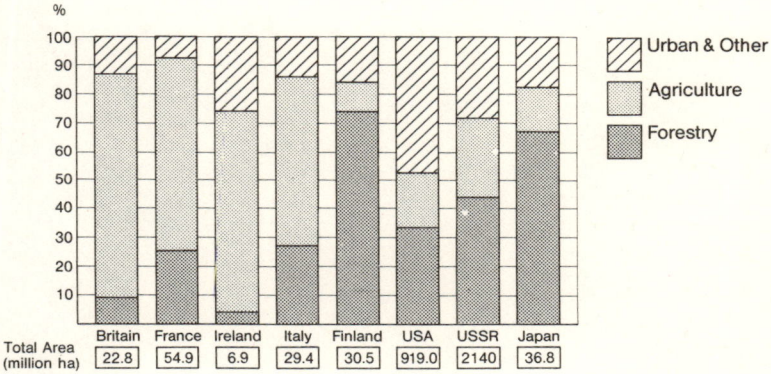

16. International Land Use: *About 80% of Britain has agricultural use, yet she is one of the world's biggest food importers, and suffers from rural depopulation and densly populated cities. Although the Forestry Commission is the largest owner of land, Britain has less forest than any other European country, except Ireland. (Source: Forestry Comm.)*

4
CONCENTRATION

"And at last," wrote John Steinbeck in *The Grapes of Wrath*, "the owner men came to the point. The tenant system won't work any more. One man on a tractor can take the place of twelve or fourteen families. Pay him a wage and take all the crop. We have to do it. We don't like to do it. But the monster's sick. Something's happened to the monster."

"But you'll kill the land with cotton."

"We know. We've got to take the cotton quick before the land dies. Then we'll sell the land. Lots of families in the East would like to own a piece of land."

The tenant men looked up alarmed. "But what'll happen to us? How'll we eat?."

"You'll have to get off the land. The ploughs'll go through the dooryard."

And now the squatting men stood up angrily. "Grampa took up the land, and he had to kill the Indians and drive them away. And Pa was born here, and he killed weeds and snakes. Then a bad year came and he had to borrow a little money. An' we was born here. There in the door — our children born here. And Pa had to borrow money. The bank owned the land then, but we stayed and we got a little bit of what we raised."

"We know that — all that. It's not us, it's the bank. A bank isn't like a man. Or an owner with fifty thousand acres, he isn't like a man either. That's the monster."

"Sure," cried the tenant men, "but it's our land. We measured it and broke it up. We were born on it, and we got killed on it, died on it. Even if it's no good, it's still ours. That's what makes it ours — being born on it, working on it, dying on it. That makes ownership, not a paper with numbers on it."

"We're sorry. It's no use. It's the monster. The bank isn't like a man."

"Yes, but the bank is only made of men."

"No, you're wrong there — quite wrong there. The bank is something else than men. It happens that every man in a bank hates what the bank does, and yet the bank does it. The bank is something more than men, I tell you. It's the monster. Men made it, but they can't control it."

How Profitable is Farming?

Along with the estate agents, banks are revelling in the over-capitalisation of farming, a viscious economic circle, a raw financial approach to the land and agriculture. The large clearing banks, including Lloyds, Barclays and the Midland, have set up agricultural departments offering special loan schemes to farmers. Banks sponsor agricultural events and conferences and send their own advisers to teach at agricultural colleges. They regard farmers as first-class, safe, clients.

Bank lending to farmers has risen sharply over the past few years, and significantly faster than lending to manufacturing industry. Loans increased by 30 percent in 1979, by 25 percent in 1980 and by 20 percent in 1981, amounting then to about £3,400 millions. But this is due to inflation, over-investment, encouraged by tax advantages, and by farmers wanting to raise cash to buy more land. Even so, farmers' total borrowings, which represent only about 10 percent of their funds, are less proportionally than those of West German, French or Danish farmers. And although of course they have a vested interest in painting a rosy, encouraging, picture of agriculture, bank advisers always suggest that farmers' complaints are unjustified. Dr Robert Bruce, agricultural general manager of the Midland Bank, said early in 1980 when the National Farmers' Union was indulging in a particularly vociferous round of complaints: "when you begin to look a little bit below the surface, a lot of farmers have no cash-flow problems, and some are doing very nicely thank you". He doubted whether more than 10 percent of farmers were in trouble. A few days later, Christopher Pettitt, agricultural finance adviser to Lloyds Bank, also played down the problems facing farmers.

While farmers are welcome, and sometimes naive, customers as far as the banks are concerned, they also benefit from substantial grants and subsidies from the Treasury; that is to say, the taxpayer. In spite of persistent claims by successive British governments that our farmers are more efficient than those elsewhere in Europe, investment assistance from the State plays a bigger role here. About 80 percent of expenditure on new farm buildings and equipment is grant-aided. The Centre for Agricultural Strategy argues that far from being held back by financial constraints, farmers buy too much machinery in profitable years purely to offset tax liabilities. Farmers enjoy total write-off for tax purposes for the purchase of machinery. "It is widely suspected," says the CAS, "that the existing structure of fiscal incentives has distorted farm investment, with excessive purchase of certain types of machinery being made to secure purely a fiscal advantage." It has called for a radical reassessment of the nature and reasons for capital assistance to farmers by the State.[1]

I will return later to the broader implications of this for agriculture and the countryside. But in the context of imme-diate financial aspects, one startling fact stands out: farmers receive more, in grants and subsidies than they pay out in capital tax. On average, they pay in tax less than 10 percent of their net income. It is generally acknowledged that an owner-occupier can get away with a total tax burden of 12 percent payable over an eight-year period though, according to Humberts, one of the leading estate agents, "the owner-occupier who takes the proper tax planning steps in good time, can avoid all or most of his tax liabilities". In a report on tenant farmers and owner-occupiers, the Milk Marketing Board, a farmers' cooperative, argued that "by using the concessions available, a farmer of reasonable management ability should be able to pass intact to his heir a farm of up to 1,000 acres". According to the Ministry of Agriculture, there are fewer than 1,000 farms bigger than this.

Farm landowners benefit from increasingly generous tax concessions which with careful planning have reached the point that in the view of many observers Capital Transfer Tax has become, like the Estate Duty it replaced, little more than a voluntary tax. By 1981 owner-occupiers could benefit from an immediate 50 percent CTT relief on business assets. By setting up partnerships, farmers can ensure that the basis for land evaluation is not the price of vacant possession but the price of tenanted land which can be up to 60 percent cheaper. If he has had a good year, the landowner can average out his income to take in the following, or previous, year. Alister Sutherland, lecturer in economics at Cambridge University, who resigned as specialist adviser to the Northfield inquiry into landownership in protest against the prevailing attitude of its members, calculates that by using other devices, including tax-deductible life insurance policies, working farmers with up to 1,000 acres could hand over the whole of their farms without any threat of fragmentation from CTT. The CAS goes further: "Even for the very large owner-occupied farms with over 1,000 acres", it says, "the evidence is that CTT in theory imposes a lesser burden than the Estate Duty provisions".[2] The Royal Commission on the Distribution of Income and Wealth in October 1979 estimated that business relief alone reduced the effective rate of tax on an estate by almost three-quarters. Even the Country Landowners' Association acknowledges that without other tax planning, the percentage lost in CTT on life-time gifts almost halved between 1975 and 1980.

The revenue raised by Estate Duty and CTT as a proportion of all Inland Revenue taxes fell from 2.4 percent in 1974/5 to 1.8 percent in 1977/8 and as a proportion of all personal wealth from 0.21% to 0.19% during the same period. The Government told the House of Commons on 9 June 1978 that the effective rates of tax for estates where half is left to the spouse and half to a son as a business was as follows:

Value of estate	Estate Duty (March 1974)	CTT (March 1978)
£ 500,000	50.6%	10.6%
£ 1 million	57.1%	12.8%

In 1981 the Government for the first time introduced a special 20 percent relief plus an interest-free instalment system for owners of tenanted land. According to Sutherland,*

*see for example, "Capital Transfer Tax and Farming" Jour. of Inst. for Fiscal Studies, March 1980.

even on assets worth as much as £4 million, the effective tax rate on personally-owned businesses will be less than 12 percent once in a generation, or about 0.4 percent a year. Even without sophisticated tax avoidance, over 99 percent of wealth owners will now be able to pay zero CTT when they hand on their assets, he says.

As the *Financial Times* put it after the 1980 budget which doubled the CTT threshold for exemption, a move which then freed from CTT at least two-thirds of the estates which would otherwise have been liable, "the declining tax take of death and gift duties may well mark a reversal of the trend which has continued in most of the country towards a reduction in equality".

Wealth, including wealth through landownership, is already spread more unequally than income. Just over 1 percent of the population owns almost 70 percent of all land. The statistics do not mean that all farmers who own land are among the richest people in Britain — and certainly they do not all enjoy the disposable income common to managers in other sectors of the economy. They do demonstrate that landownership is a special privilege in terms of wealth. And though there is a stream of individual large estates having to be sold off by their owners (Lord Rosebery's Mentmore Towers, Lord Egremont's Wastwater, the Duke of Argyll's Iona) what is much more striking when it comes to painting the overall picture is that landed wealth has not been redistributed to any significant extent. Most wealth is still inherited and inheritance of land — as indeed, of other assets as well — is the most important single cause of inequality in the distribution of wealth.

The tax system carries implications not only for wealth, but also for agricultural policy and the kind of farming system and even the kind of countryside we are going to have. Tax concessions are encouraging the growth of large farms, benefitting owners of land well over the average size of holding; this is well above the 200-300 acres that economists, let alone ecologists, agree to be beyond which economies of scale generally lead to diminishing returns. The concessions also promote increases in the price of land which, in turn, lead to higher costs and higher food prices.

The one use governments in the past have made of tax as an

instrument of coherent policy is that of discriminating against landlords — owners of let land as opposed to owner-occupiers. The CLA argues that as a result damage is done to agriculture by landlords being forced to abandon investment plans or protect parkland, which landlords have to pay for by putting money aside in anticipation of a future tax burden. 80 percent of let land is still in private hands, but landlords say that tax pressures will mean that they will be forced to farm more of their farms themselves, rather than letting it out to tenants.

The trend to owner-occupation is being encouraged by the Agriculture (Miscellaneous Provisions) Act 1976 which strengthened tenants' security by enabling tenant farmers to hand down their holdings for up to three generations. Landowners agree that this constraint will ensure that as soon as they have the opportunity to take land, previously rented, in hand they will seize it. They are prepared to rent, but to whom they want and when. They also insist that the landlord-tenant system is, by definition, a good thing for agriculture, an assumption which should not go unchallenged. It can be argued equally well that owner-occupiers will care for the land better than tenants and, in any event, a very small percentage of lettings go to new entrants to farming.

Farmers argue, meanwhile, that conventional measurements of their wealth, the value of their land at current market prices, are inappropriate since their assets are fixed, a fact that gives rise to the adage 'the only rich farmer is a dead farmer'. A report from the 1978 Cambridge Agricultural Policy Conference said: "there is only one way in which a farmer can get some spendable benefit from the ownership of very valuable assets and that is by running up an overdraft upon the security of the farm and hoping that inflation will make the borrowing eventually worthless". Hence farmers' propensity to go to the banks.

A more sophisticated argument comes from Donald Denman, former professor of Land Economy at Cambridge University. He says that by accepting yields below the value of their capital inputs when compared to industry farmland owners are providing what he calls a 'private subsidy' to agriculture. Private landlords are prepared to do this, Denman says, because of the many other advantages of agricultural land ownership: a pleasant residence, the principle of inheritence, the feeling of social responsibility, the enjoyment of amenity, love of the countryside, and use of allied resources such as timber, materials and sporting rights. Though he avoids the question of who (the taxpayer) contributes through grants and subsidies for many of the capital inputs, he says that many landowners can provide this 'private' subsidy since they have non-agricultural wealth equal to their assets in land.

But it should not be forgotten that taxpayers and consumers also subsidise farmers heavily through tax allowances, high EEC guaranteed farm prices, and the non-rating of farmland. The Government also exempts landowners from tax if they can convince it that the property is part of the national heritage according to the 1977 Capital Taxation and the National Heritage guidelines. One notorious example was the Treasury's decision, after consultation with the Countryside Commission and the Nature Conservancy Council, that the Vestey family's 3,000 acre estate in Sutherland was "of outstanding scenic, historic, or scientific interest". If there had

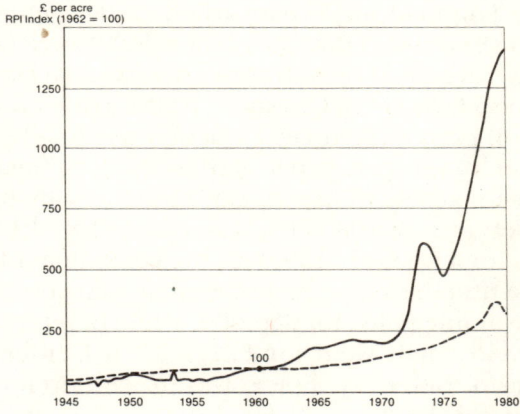

17. The Price of Land: *Farmland is an unique asset. As Mark Twain said "They don't make it any more". At times of economic crisis it is a secure investment and a hedge against inflation. High land prices are one of the main obstacles to entry into farming and makes agriculture virtually a closed shop. (Source: M.A.F.F.)*

been no exemption, CTT charges would have amounted to £540,000.

By forming a partnership or a company, landowners can pay the lower CTT burden based on the value of shares rather than on the value of the property. And landowners can divide up their holdings so as to be able to claim more grants from the government or the EEC (grants are paid out to individual farms rather than individual owners) just as they can divide up their holdings to benefit from the highest rates of tax relief. The possibility of enlarging and then splitting up holdings to suit different tax brackets are immense. So established farmers are tempted to buy up neighbouring land at marginal extra running costs even though their earnings from the extra land do not justify the price they paid for the 'new' land. The ability and willingness of farmers to participate actively on the market must be the result of high expectations of income, increases in the value of land and prices despite continuing complaints about incomes.

This was well illustrated in 1978 when the Somerset family of Jeanes, acting jointly as Stowey Farmers Ltd., bought the 2,500 acre Cricket Malherbie estate for £2 millions from the Beaverbrook Foundation. A year earlier, estate agents Wright and Partners announced that they had sold 30-acre blocks of land at £1,600 an acre, then well above the national average, to a group of local farmers in Cheshire. At the same time, other farmers sell off parts of their land; but because small plots can be sold at such high prices per acre, only those farmers or developers with access to capital can afford to buy it.

A Somerset farm of 108 acres was sold in July 1979 in 15 separate lots for a total of £459,000, the equivalent of £4,252 an acre. One tithe barn (for improvement) was sold for over £26,000, two arable fields for £60,000, while barely 2 acres of land for development went for £72,000. In October 1980, Southend Farm, Billericay, Essex, was broken up into seven fields selling for between £1,500 and £5,300 an acre to neighbouring landowners.

The entry into the market, especially in good quality land, by City institutions, has contributed to the rise in the price of farmland over the past decade. But farmers themselves contributed to this rise, specifically through the 'roll over' provisions of the 1972 Finance Act which enabled landowners who sold land for development to avoid Capital Gains Tax if they reinvested their profits in agricultural land within twelve months. Many farmers were encouraged to pay well above current market prices.

Farm rents are also rising rapidly, stimulated in part by the ability of financial institutions to pick and choose their land-hungry tenants and charge them rack rents, and by concern among farmers about the need to gain succession rights for their sons. Farm rents more than doubled between 1970 and 1976 and have increased on average 20 percent a year since then. They are reviewed every three years.

Tenancy legislation, the price of land, the capital-intensive nature of modern agriculture and the abuse and sale of local authority smallholdings have all contributed to the obstacles in the way of an entry into farming. Farming is a closed shop.

Changes in Farming

The debate about the actual wealth of landowners will continue — there is, after all, a distinction between fixed capital assets and income — but there is no doubt at all that it is extremely difficult to become a farmer. The farming ladder has been broken at the bottom. The Northfield Committee described how some of those giving evidence to it argued that "it would not be desirable if entry to farming were restricted solely to a privileged class of inheritors or to those few with large sums of capital to buy themselves in". The only option open to the 60 percent of the 700 or so agricultural graduates turned out by universities and colleges every year who are not sons of farmers is to become farm managers or join large agribusiness corporations. (And it is significant, indeed alarming, that increasingly agricultural research at universities is being sponsored by private companies.)

There are between 1,000 and 1,500 openings for new entrants each year, but this includes jobs as managers and those already farming but wishing to move up the ladder. No more than a third of new lettings go to farmers setting up for the first time. More than 100 people, on average, seek applications for every tenancy and partnership offered by institutions. In 1979, more than 200 individuals sought the 1,000 acre tenancy offered by Eastbourne borough council. A comprehensive survey carried out by Manchester University showed that to start a small farm of 125 acres, you would have needed £300,000 in 1979 compared to £33,000 in 1971. "Opportunities to farm, except for farming families, are few and likely to decrease further", the Northfield Committee concluded. And the financial pressures on small farms, the move towards larger and more specialised holdings, policies which make rich farmers richer and the poor, relatively poorer increasingly threaten existing family farms.

Can Britain learn from the ways other countries have tried to tackle the problem? In France, where the problem is that too few young people want to enter farming — especially in hill areas — the government tries to encourage them by offering heavily subsidised credit and grants. Up to 90 percent of the cost of buying land there can be covered by loans from the Credit Agricole (the world's largest bank). Control over land purchases by the SAFERS (see chapter 2) helps to make land available, while the law on the accumulation of farm businesses is designed to limit the size of a farm to that worked by one family. In Ireland, the Land Commission tries to perform functions similar to the SAFERS. In Denmark and the Netherlands, where there is severe pressure on a limited land area, loans are cheaper and more readily available than in Britain. In New Zealand, share-farming especially in the milk sector has helped young farmers. Under this system, a kind of partnership, the landowner and established farmer provides the land and the share-farmer provides a proportion of the livestock and working capital. Profits are shared according to the proportion of assets owned by the different partners. But the problem here is that there is always going to be a senior, dominant, partner. Not surprisingly, the practice has caused tremendous bitterness and controversy.

In Britain, we have seen how farms have formed partnerships and companies (which now probably account for about 40 percent of all farms and most farmland) and trusts. They close up landownership or farming to a small group. They encourage capital accumulation and increases in the price of land and in the size of farms, making it even more difficult for the outsider, or newcomer, to enter agriculture.

(One technical way into farming, though not landownership is contract farming. Under this sytem, which is becoming more and more widespread in Britain, specialists — dairymen, for instance, or sheep shearers — hire out theirlabour to farmlandowners.)

As agriculture, in the last resort the most important sector of the economy, becomes increasingly capital — and energy — intensive, it employs fewer and fewer hands. "For most of this century," as the National Council of Social Service has put it, "farm structure has basically been determined by economics and technology, and not by the real needs of the land and farming people."

As the number of farms is falling by about 5 percent a year;

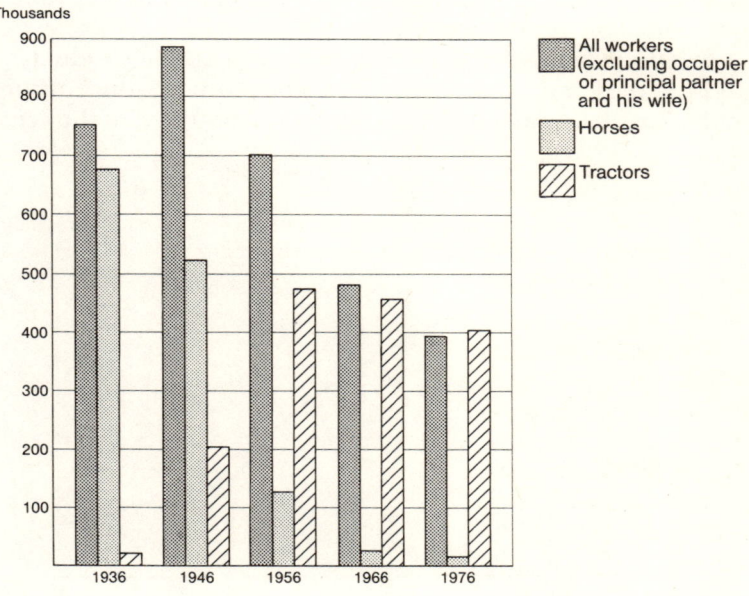

18. **Labour:** *Less than 3% of the active population of Britain work on the land, by far the lowest proportion in Europe. There are now more farmers than farmworkers. The apparent lower cost of machines compared to people fails to take into account the long-term issues of health and fertility. (Source: M.A.F.F.).*

the average size increases. The number of small farms, those less than 100 acres, has been halved over the past ten years.

About 1,500 farms disappear every year. The average farm size is now about 120 acres. But this disguises the fact that food production is fast coming under the control of a very few large landowners. The largest ten percent of farms account for half of Britain's food production. Farms under 500 acres account for less than 6 percent of the total acreage.

As farms are getting bigger, so they are becoming increasingly specialised, a world away from the traditional picture and rational structure of mixed farms with corn, chickens, ducks and cattle, that legacy conjured up now in cartoons, children's books and advertisements for food. For instance, a little over 6 percent of specialised pig farmers account for half of Britain's production of pork and bacon. Two-thirds of the country's total egg production is in the hands of less than 2 percent of all producers.

Britain is being ruthlessly divided, with arable products — and the pigs and poultry which eat the grain — concentrated in the East, and dairy and beef cattle concentrated in the West,

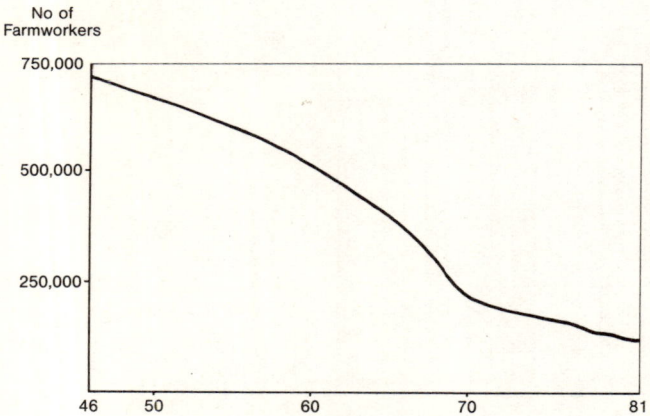

19. Number of Farmworkers: *With the notable exception of the war years, the number of farmworkers has fallen consistently since the number of industrial workers overtook those on the land in 1851. There are now more farms and more tractors than farmworkers. (Source: M.A.F.F.)*

where heavier rainfall helps the grass grow thicker (See fig. 12, p69) During the 1976 drought, many lorryloads of hay had to be transported from eastern counties to the West and Wales to feed livestock.

The move towards bigger and bigger units is not unique to agriculture. But British politicians and the farming establishment are particularly proud of agriculture's record. They point to the fact that productivity in agriculture has increased about twice as fast as in industry. Yet productivity is measured solely in terms of the number of people who work on the land, not in terms of the social, economic or environmental costs.

Over the past twenty years the number of people in the EEC earning their living from the land has dropped from 17 million to 7½ million. The French government is concerned already about the effects of rural depopulation, but the British is proud that here one person in farming feeds 42 people, while in France the figure is 'still' less than 25. In Britain, at the end of a period of twenty years during which the number of farmworkers halved, with a smaller proportion working on the land than in any other European country and the 140,000 or so who remained, close to the bottom of the wages league (though they are employed by some of the wealthiest individuals in Britain), Dr Keith Dexter, the Ministry of Agriculture's chief farming policy adviser, told students at the Royal Agricultural College at Cirencester in 1978 that society would have to pay the social costs of new technology. Redundancy payments, he said, should be paid to those working on the land "whose production is no longer required".

Yet Dexter himself admits that whatever the type of farming, the family farm of two-to-five men can carry out all the production activities of a large farm and that many large farms are divided into smaller units for easier management. According to a 1961 report, "Scale of Enterprise in Farming", most of the gains in efficiency occur up to 100 acres, and very few benefits of increase size occur beyond 200 acres. Another major study[3] concludes that economies of scale ran out at the 250 acre limit for dairying, 400 acres for cropping, and that there was no consistent increase in efficiency beyond 400 acres. Indeed, farms produce a consistently lower value of output per acre than comparable medium-sized farms. The

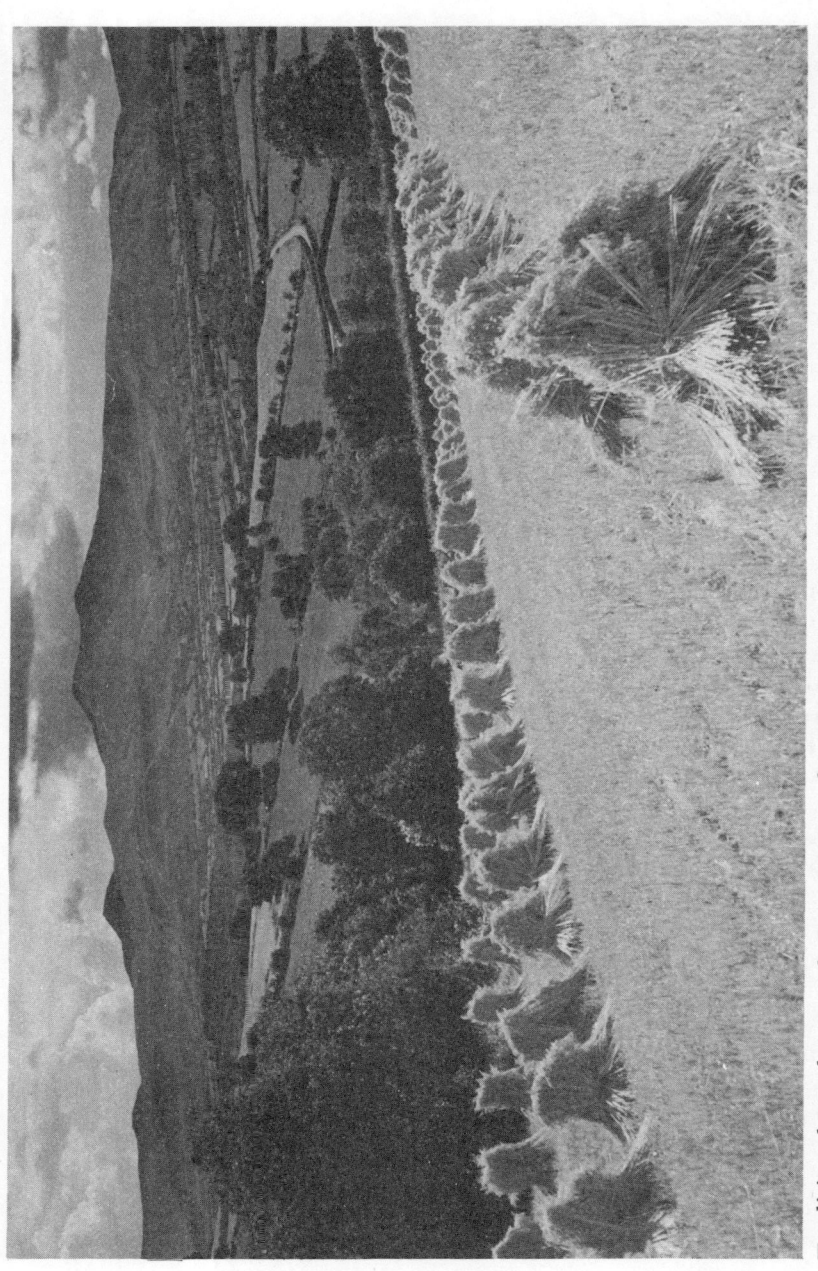

Traditional Landscape. *The Brecon Beacons from Battle Hill, a scene showing how traditional and mixed farming can blend with the landscape. How long will the trees remain, and the Beacons remain uncovered?* (Photo: Wm. Meadows)

report concluded: "we have found very little evidence that any appreciable economies of size are generally to be gained by enlarging beyond the point where the farmer remains in very close contact with his employees, if indeed he has any at all outside the family . . . We are impressed by the frequency with which American studies affirm that, in most farming situations, all the economies of size can be achieved by modern and fully-mechanised two-man farms, and that even one-man farms can attain highly efficient systems of operation".

Small farms help to prevent rural depopulation, they allow more entrants into farming, they reduce dependence on large and distant companies, absentee landlords. If small farmers amalgamated in the search even for average efficiency, this would have a tiny impact on food production, according to conventional criteria.

Yet most small farms are not eligible for Ministry of Agriculture grant schemes. Small farms are neglected by the Government and researchers. Policies are formulated — or simply allowed to take their course — without any thought about the impact on small farms. They are ignored by the Ministry for whom big equals the most efficient, the producer of the cheapest food. Big farms for the bureaucracy are simpler to deal with and more compatible with the priorities and development of the Common Market's Agricultural Policy.[4]

British policy-makers equate size, that is to say, bigness, with efficiency. How can that be? The average size of farm-holdings in Britain, and of livestock herds, is well above those on the Continent. Yet during the autumn of 1980, when British farmers were benefitting from the highest prices in the EEC because of the relationship between the strong pound and the Green Pound used for intra-Community agricultural trade, the NFU said that these prices were insufficient to cover their costs. The Centre for Agricultural Strategy has pointed out what independent analysts had been dismissed out of hand for saying before; namely, that on the basis of comparative costs, British farming is not as efficient as farming elsewhere and notably the Netherlands, Belgium and Denmark. It says that there was a case for expanding British agriculture, but that this should be achieved by a less wasteful use of capital and public money and a more constructive approach towards labour.

Concentration and specialisation has also brought centralisation. It is scarcely efficient to trade vegetables through the new Covent Garden market (Nine Elms) in London before sending them back to consumers who live a few miles away from where they were grown in the first place. Large plant bakeries not only produce poor-quality bread filled with chemicals to enable their loaves to travel and rest for days on shelves; the cost of transporting the bread puts up the price. Associated British foods (makers of Sunblest) and Rank Hovis McDougall (producers of Mother's Pride) acknowledge that their overheads and costs are much higher than those of local, hot bread, shops. Food — once the basis indeed the origin of local markets — now accounts for more road traffic than any other product.

Those farmers, in the meantime, who have increased their acreage have not become more independent. Many are trapped in a vicious circle: the higher their costs, the more they try to produce; but they are encouraged to believe that the surest way to produce more, so as to cover their costs, is to rely more on (increasingly expensive) chemicals and machines. Day after day, salesmen from chemical companies, seed companies, machinery companies, processing companies, knock on their door, offering the latest product on 'special' terms that may include discounts, long-term contracts coupled with 'easy' credit terms and other seductive package deals. Farmers are made to feel that they must keep up with the latest in modern farming methods, giving the quickest, easiest and highest yield. And so once again the long-term interest and health of the land and the soil is pushed well into the background.

Massey-Ferguson (the world's biggest farm machinery producer which recently laid off thousands of employees) in 1978 launched a new direct seed drill, to be used specifically with paraquat, the highly toxic herbicide, produced by ICI. It insisted that for Europe's farmers, "a key advantage of direct drilling is that by eliminating ploughing and cultivation work, it dramatically cuts the time needed to establish a new crop in the busy post-harvest period when the weather frequently limits the number of days available for field work".

Farmers are becoming increasingly dependent on contracts with large food processing companies which not only impose a price on the grower, but also tells him what varieties of individual crops to sow, how to cultivate them, even when to pick them. More than a thousand farmers in East Anglia sign contracts every year with Unilever's large Bird's Eye freezing factory in Great Yarmouth. Then the company decided to close the plant. As one neighbouring farmer and director of one of the few cooperatives in Britain put it: "The major processing companies have higher overheads — sophisticated freezing and packing plants, investment in research and development and a larger expenditure on advertising. To cover their costs, they are moving into added-value packs such as potato croquettes, which command a higher profit margin".

Food Monopolies

Farmers do not help themselves; they are not effective at marketing their own products, and indeed for the most part they are not bothered about what happens to their food once it leaves the farm gate. They know that large companies are secure clients, since more than 90 percent of all the food we eat is already processed one way or another with the help of 2,500 registered additives. Far too often farmers seem oblivious of the fact that it is now the food industry, not agriculture, which dominates the food chain.

The West is saturated with food and obesity has become a national disease (poverty, not a shortage of supplies, is the cause of hunger). Faced with a stagnant market, companies in order to compete have little choice between taking over competitors or be taken over by them. The market in sugar, bread, milk and eggs, four staple foods, is dominated by a monopoly or near-monopoly. More and more companies are squaring the circle by taking an interest in every link in the chain: farming, milling, processing, supermarketing, research and development.

Unilever owns BOCM Silcock, Britain's biggest animal feed manufacturer, Bachelor Foods, Bird's Eye, Walls, East Sussex

Farmers Ltd., Dale Turkeys, Matteson's hams, Flora and Blue Band margarine. It controls more than two-thirds of Britain's margarine market. In 1972, the company chairman, E. Woodruffe, said: "If our profits go on growing at the rate they did last year we would wipe out the world".

Seven years later, Unilever sought and obtained permission from the Office of Fair Trading to sell its Macmarket shops to British American Tobacco Industries, Britain's largest cigarette producer. BAT, which owns the International Stores chain, gave Unilever some of its share capital in return. Another giant conglomerate, the Imperial Group, manufacturers of Players and Wills, diversified into food, and is now Britain's largest poultry farmer and one of Europe's biggest egg producers. It controls Courage's brewery, Smedley Foods, HP Sauce, Lea and Perrin's Worcestershire sauce, Young's Seafood, Ross Foods, Buxted Poultry, Golden Wonder crisps. Both Unilever and Imperial own trawler companies based in Humberside. In 1980, the Office of Fair Trading was concerned enough to conduct an inquiry into the operations of Daylay Eggs, an Imperial subsidiary.

Of the 90 brewing companies in Britain, six control 80 percent of the market and own most of the tied pubs. They are: Bass Charrington, Allied Breweries, Courage, Scottish and Newcastle, Whitbread, Watney-Mann (part of the Grand Metropolitan Group which also owns Express Dairies), and Truman. Allied Breweries took over the John Lyons tea and food group in 1978 — a year later it bought its third farm, in Herefordshire, "to safeguard its own hop supplies", as the *Farmers' Weekly* put it.

Barely twenty years ago, there was no national brewery chain. But when the brewing companies merged, the new, larger, company closed down many pubs, and replaced tenants with managers. They also promoted keg. Just as chemicals are added to bread to enable it to travel well and prolong its shelf-life, so the advantage to the brewers of keg over traditional beer fermented in casks is that keg is chilled, filtered and pasteurised in the brewery where the fermentation stops. It thus needs little care and is easily transported from central plants. (The success of the Campaign for Real Ale speaks for itself. Allied Breweries' Ind Coope, for instance, has decided to split up its chain of pubs in London and bring back such traditional names as Taylor Walker. More and more pubs have gone back to serving draught beers.)

Two-thirds of the £400 millions worth of tea drunk in Britain each year is packed by three companies. A third of the money spent on food in Britain goes into the cash registers of just six supermarket chains. In addition to its supermarkets, Sainsbury owns one of the largest pork processing plants, jointly with Dalgety owns an egg and poultry packing station, and jointly with Pauls and Whites' animal feed manufacturers owns Breckland Farms Ltd., a large intensive pig producer in East Anglia. It also owns two beef farms in Ayrshire.

The Dewhurst butcher chain is owned by the huge Vestey meat empire, with ranches in Australia, South America, New Zealand and Scotland. Barely two years after Spillers opted out of its loss-making bread-baking activities in 1978, Dalgety, a large good conglomerate which also owns cattle stations in Australia and New Zealand and livestock farms in Scotland, took over the rest of Spillers' interests. In the course of its aggressive campaign, including a series of full-page advertisements in national newspapers, to persuade Spillers' shareholders to accept its bid, Dalgety stated proudly:

> one pint of beer in every six is made with our malt,
> one supermarket egg in every twenty is ours,
> one pig in every twelve is bred from Dalgety stock.

Over the previous five years, another advertisement pointed out, Dalgety's profits had risen by 363 percent. Less than half the company's turnover is in Britain — the rest is in Australia, New Zealand and America. It has interests in cereals, malt, animal feed, timber, wool, frozen vegetables, chemicals and meat. "Thus", the company says, "we can ride out a storm in any one continent or product group, secure in our profits from the others". When it took over Spillers, Dalgety said that the combined group would be substantial in grain, animal feed, malt, petfoods, and flour — "certainly able to handle competition from anywhere in the world". Dalgety wanted to take over Spillers so that it could offer farmers a 'complete package': selling chemicals, seeds, fertilizers and breeding stock, and market the produce throughout the world. Some

food companies, including Dalgety and RHM, are also grain
brokers. they know the market and can influence it to suit their
own needs, products and profits. In comparison with the
information the companies have at their disposal, the
individual farmer is ignorant, virtually helpless.

In another example of the tight hold of a few companies
over the market, the erstwhile Price Commission in 1978
strongly criticised the lack of competition in the £1,200
millions a year animal feed industry, dominated by Unilever.
Six out of seven companies which accounted for over half the
market co-ordinated their price increases throughout 1976, it
said. It noted that BOCM Silcock's cash profits more than
doubled over a period of five years and that the company had
introduced a system of 'loyalty discounts' to customers. The
Price Commission also attacked the companies for not providing adequate information about the contents of their increasingly sophisticated products. Many farmers were concerned, it
said, about the thought that they could be paying £100 a
tonne for compounds containing unknown quantities of
materials "such as straw and dried poultry waste".

When Spillers opted out of bread-making in 1978 to
concentrate more on its profitable flour-milling operations
and overseas investments, it left RHM and Allied Bakeries, a
subsidiary of Associated British Foods (ABF), in control of
about 60 percent of the bread market. More significantly,
RHM, Dalgety and ABF now control about 80 percent of the
flour used in Britain. ABF is part of the world-wide empire of
the Weston family which came to Britain from its native
Canada and invested directly here partly to encourage British
bakers to buy his Canadian flour. It owns the Fine Fare chain
of supermarkets. As well as grinding its own flour, RHM also
packs butter. makes Saxo salt, owns Atora suet, Energen rolls,
Vencat curry powder, Sherwood's chutney and spices, Bisto,
and Scott's porridge oats.

We shall return to the story of British bread. But in the
whole area of staple foods, successive, and particularly Labour
governments. have openly admitted to having turned a blind
eye to the letter and spirit of monopolies legislation, mainly in
the naive hope that large conglomerates will be more able to
keep down the price of food and maintain the labour force.
This hope is a forlorn one.

In the United States, 85 percent of the land is owned by 12 percent of the farming companies. These include such large corporate interests as Shell, South Pacific Railroad and the Bank of America. In Britain, Peter Walker, Mrs Thatcher's Agriculture Minister, proudly pointed out, 30 firms control half the food manufacturing industry and 50 firms control half the total food distribution sector.

The ultimate and, for as far as the companies are concerned, the most logical move towards total control of the market is their little-noticed take-over of the trade in seeds, a market worth more than £5,000 millions a year, if the attractive and potentially highly profitable Third World market is taken into account. In the space of a few years, Royal Dutch Shell, the world's largest company, has become the world's largest seed company. It has secured control of Nickerson's Seeds and the Norfolk-based company, Farm Seeds. These large corporations are concentrating on the profitable hybrid, high-yielding, and uniform varieties that are vulnerable to disease and dependent on high applications of fertiliser. It is no accident that the companies involved are also chemical manufacturers and drug producers: many drugs are by-products of plants. Shell is now an important carrot farmer in California, while Mobil Oil company is busy producing tomatoes. Other agrochemical companies which have joined Shell in their search for control of the seed market include Ciba-Geigy and Sandoz of Switzerland and Pfizer and Union Carbide of the US. The companies have also been encouraged by seed patent legislation which both protects their own varieties from competition and guarantees them royalties from farmers. It was this which encouraged RHM to buy up 84 small seed merchants in one week in the 1970s. By using hybrid seeds, which do not often reproduce, farmers have to return to their supplier each year; they cannot follow the traditional practice of keeping part of the crop for sowing for the next season's harvest.

The supreme irony is that large companies have now turned full circle: Booker, McConnell has invested heavily in health foods, owning the Holland and Barrett chain and Allinson's stone-ground flour. ABF and Sainsbury have set up their own 'hot bread' shops.

Although, as we have seen, these large companies do own farms, they are not as vulnerable as farmers to the risks of falling prices, for instance, or bad weather, or government policies. If the price of palm oil in Malaysia increases, Unilever can turn to other suppliers or even to substitutes. Nestlé, the world's second largest food corporation, does not own a single farm. With a finger in many different pies — some of them totally unconnected with agriculture — and in every part of the food chain, big companies can insure themselves against short-term risks and, with their widespread interests, they can take short-term advantage from virtually any situation. With a strong capital base behind them, they can survive the increasingly capital-intensive food producing system and the High Street price wars which perpetuate the traditional assumptions shared still by politicians and the majority of urban consumers — namely, food should be cheap above all.

The companies, and indirectly us as consumers, encourage farmers to go for the fastest possible yield, no matter what damage in the meantime is done to the land. They widen the gulf between town and country, however much they insist on perpetuating the 'wholesome' and thoroughly bucolic and outdate image of farming on their plastic packets and their tins.

Large food companies are tightening their hold over the farming community and agricultural policy. But it must also be a savage indictment of the system when farmers in Britain, enjoying higher prices than farmers anywhere else in the EEC, with the largest holdings in Europe and with those employing full-time labour earning an average income of £20,000 a year, and with tax bills proportionately less than their employees, when farmlandowners say that they cannot afford to increase the pay of farmworkers by more than a handful of pounds a week, and when the pay of farmworkers are still close to the bottom of the wages table.

5

POLICY FOR HEALTH

"Wise consumption is a far more difficult art than wise production" – *John Ruskin.*

Every year, at the beginning of July, the farming establishment congregates at the Royal Show in Stoneleigh, Warwickshire, for its annual celebrations. But the rows of scrubbed cattle, shining machinery, bowler-hatted judges and smooth-talking salesmen perpetuate an air of confidence and complacency which belies the reality of the world outside.

They disguise a host of basic, yet unanswered questions. Chief among them: does the government have a coherent policy towards agriculture, the nation's health, the pattern of landownership, rural employment? The easy way has been chosen, but where has it led us? to fewer farms with higher costs; poor farmworkers and expensive machines; bigger yields and more surpluses; rich cereal barons and poor livestock farmers; the production of more and more food we do not want or the consumption of which we are advised to cut down; and almost as dependent as ever on imports.

It has penalised those traditional landlords whose priorities are good husbandry and care for the countryside and encouraged and helped those enterprising businessmen whose main motive is profit and intensive production, no matter the cost to the land or the environment.

The questions are not answered: they are rarely asked. Little or no considered thought is given to the role of agriculture in the overall economy, or the role of rural areas in the context of a policy towards the regions (where priority has consistently been given to propping up depressed industrial areas). This is in stark contrast to the debate in other countries in Europe. In

France and West Germany, agriculture is given a special place in the economy; in Norway, the government, following a long and animated debate in Parliament, adopted a national policy towards nutrition. We have noted that Whitehall persistently argues that British farming is the most efficient in Europe and attacks West German agriculture for its inefficiency. Yet West Germany, with a growing number of part-time farmers (now about half of all of them) produces more food than Britain on less cultivated land.

Agricultural Policy

The one consistent theme in agricultural policy-making is the political strength of the farming lobby and the way farmers are protected. But while the National Farmers' Union and the Country Landowners' Association together are an extremely successful and influential lobby, their aims seem to be limited chiefly to securing the biggest possible capital grants and tax reliefs, and highest possible prices. It is a crude approach, a reaction to Britain's traditional 'cheap food' policy of relying on supplies from America and the Commonwealth — trading links, established during the period of Empire, which have hung round the necks of those responsible for shaping a proper agricultural policy for Britain. Agriculture has been taken for granted; the urban population has little understanding of farming; it has been encouraged to rely on supplies of cheap food. So long as the supermarket shelves are full, no questions are asked. And as the produce becomes more highly processed and the search for fast, convenient, food continues apace, so the gap between town and country widens, and the relationship between consumers and the land more and more tenuous.

Over the past 150 years British agriculture has been jolted by four main events. The repeal of the Corn Laws in 1846 was a dramatic — and, for the Tory party, a politically painful — break with protectionism and led to supplies of cheap food primarily from North America, for the rapidly growing industrial centres in Britain. Agriculture was subjected to the demands of other sectors of the economy. The Second World War then forced Britain to depend more on its own food supplies and become as self-reliant as possible. After the war Labour's 1947 Agriculture Act, which gave farmers a system of minimum prices, but still kept the door open to traditional overseas suppliers, was a recognition that agriculture needed a secure basis, namely price guarantees. It was a recognition, too, that Britain needed to grow a minimum of its own food requirements (the policy during the 1930s consisted merely of introducing panic measures in an attempt to protect agriculture from the depression). Then 1973 Britain finally abandoned its policy of free trade by joining the Common Market's protectionist, high-priced system.

The experience of the Second World War is particularly significant. With 800,000 fewer acres, requisitioned from agriculture for military and other purposes, food production expressed in terms of calories almost doubled, and in terms of protein more than doubled. By 1944, wheat production had increased by 90 percent over its pre-war level and vegetable production increased by almost 50 percent. The fat content of the nation's diet fell by six percent, while the protein intake increased by the same amount.

Nearly 6 million acres of grassland were ploughed up and a number of specific measures, notably a slag and lime subsidy, were introduced to improve the quality of the soil. The amount of wheat fed to livestock was cut by half and the amount of expensive, imported, animal feed concentrates was inevitably cut dramatically. Milk yields fell, but 20 percent of the total acreage was saved by reducing the role of livestock as a source of energy. While the number of pigs and poultry fell sharply since they competed directly with people for grain, the number of cows actually increased, though their milk yield fell.

Over half of all manual workers kept either allotments or vegetable gardens.[1] In 1941 food imports were only two-thirds of their pre-war level. Imported animal feed fell from 8.7 million tons to 1.3 million tons in 1943/44. Pig Clubs were started up throughout the country. The Ministry of Food urged people on through the 'Dig for Victory' campaign (can civil servants now inspire the public?) and reported that "one acre of wheat saved at least as much shipping space as seven acres of best grass".

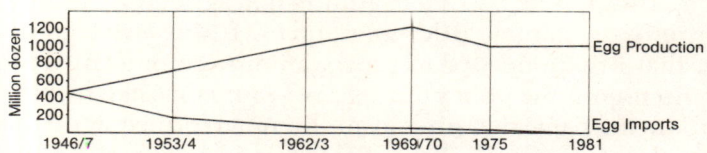

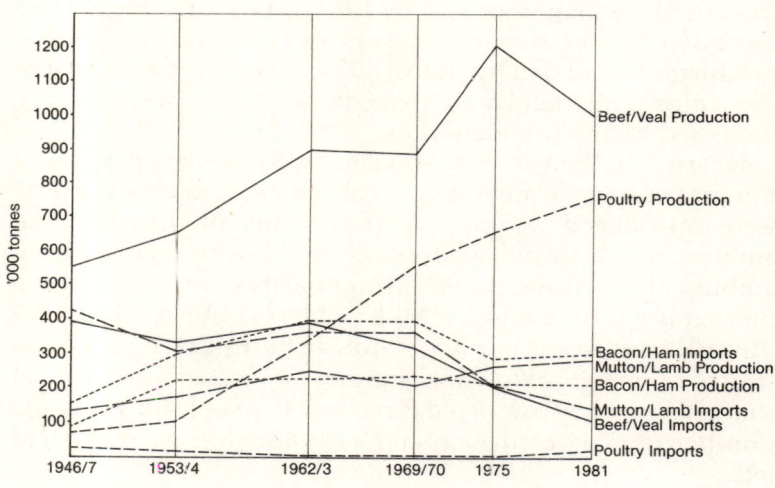

20. **Trends in Agricultural Production:** *The production of wheat and barley has increased significantly over the past fifteen years, encouraged by high and guaranteed EEC prices, the development of new varieties, and by mechanisation. But much of it is used only for animal feed or is destined for*

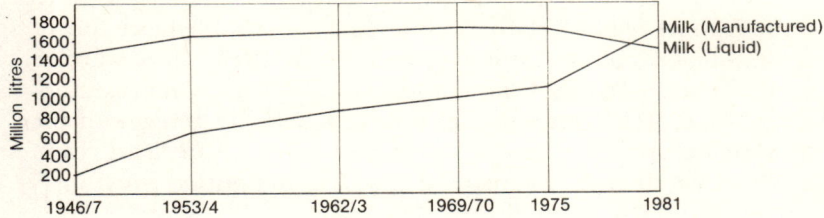

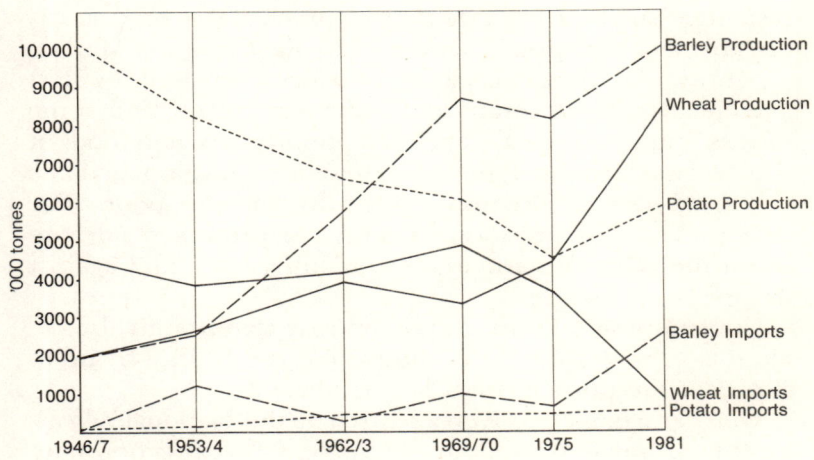

export. The EEC's open-ended system of price guarantees has led to over-production and to encourage by intensive rearing techniques offering a high protein product and a cheap alternative to red meat. (Sources: M.A.F.F. & Milk Marketing Bd.)

The number of allotments increased from 815,000 in 1939 to 1,400,000 in 1943, a year that produced a record harvest. Whitehall, which over-estimated the volume of food imports that could come in by convoy, was surprised and relieved. In 1939, there were about 12 million acres of arable land and 17 million acres of permanent grassland. By 1944, there were 18 million acres of arable land and 11 million acres of grassland. The total arable area had almost reached the acreage of the 1860s before Britain began to rely on cheap American grain. Wheat yields rose because the land had benefitted previously from the fertility given to it by grass.

Crops were grown on the Windsor Great Park which became the largest wheat field in Britain. Crops were grown on the Sussex Downs for the first time since Saxon days. They are being ploughed up now for quite different reasons: not because of the nation's need but because the agricultural system persuades farmers to grow as much arable products as possible at a high price guaranteed by the Common Market.

Meanwhile, the acreage of potatoes increased by 100 percent. Roast cormorant and eagle were offered on some menus, but the diet was generally monotonous, and food, of course, was strictly rationed. However nutrition actually improved during the war, especially for the poor. The Government took nutrition seriously and promoted a debate about the value of basic foods, including offal, and against waste.

Consumption of milk and bread rose significantly, but so did that of vegetables — by about 30%, and fruit — by up to 50%. Consumption of meat fell significantly.

One particular controversy centred on the 'National Wheat-meal Loaf' made from flour of a high, 85%, extraction rate. Though unpopular, it was far more nutritious than what has become the traditional English sliced white loaf, simply because more of the wheat was used.

So people were adequately fed, indeed more than adequately. Yet this was a response to an emergency; it was not the result of a thought-out policy. The attention paid to nutrition was regarded in the main merely as an economic necessity: fat consumption soared after rationing ended in 1952.

The demands of the war persuaded farmers to use straw rather than waste it; sometimes it was pulped for cattle feed, sometimes it was sold for paper. Farm incomes soared — farmers benefit financially from crises — but the war also left Britain with the most mechanised agriculture per acre in the world. The war, too, showed how a government could control farming, and monitor every single farm, through the network of the County War Agricultural Executive Committees, though since these consisted mainly of farmers, landowners were judged by their peers.

The switch to an entirely different agricultural policy and perspective came in 1973. The decision to join the Common Market brought high tariffs on imports of food from outside the EEC. British farmers could look forward to much higher prices than those prevailing on the world market and than those the Treasury previously guaranteed them.

The decision was taken without any coherent, articulate debate about what was best for British agriculture or British consumers. It is significant that the NFU was traditionally opposed to Common Market entry but was finally seduced by Whitehall. The Ministry of Agriculture adopted the view that all it would need to do once Britain was a member was to sit back, comforted in the knowledge that its client farmers would be kept happy with the prospect of their prices rising substantially from the relatively-low British levels to the high EEC levels.

It turned out to be not quite as easy as that. Over two-thirds of the entire EEC budget is devoted to agriculture, and to large food processors and traders in particular. The budget is made up mainly of import duties on industrial and agricultural products imported by EEC countries from the rest of the world. Britain is still one of the world's largest importers; at the same time, it has a smaller number of farmers, relative to the rest of the active population, than other member countries. Britain, therefore, under this system is one of the largest net contributors to the EEC budget. In addition, consumers have to pay much more for their food than they did in the past (and I am not suggesting that food should be cheap, simply that a sensible food and farm policy should be discussed and implemented). According to official Whitehall figures, the

EEC's common agricultural policy has increased Britain's total food bill by at least £2,250 millions a year. Most Ministers and civil servants try to play down the problem.

Others present it in crude terms. For example, Roy Hattersley, then Labour's Prices Secretary, said in November 1977: "Our farmers are historically efficient. We became so suddenly, when the Enclosures filled the factories and amalgamated the small farms and began the long tradition of British agricultural excellence . . . The Common Market will have to buy out thousands of inefficient farmers. Until it does, we shall not have an efficient farming policy. Until we have, the EEC will remain at its present level of severe and potentially disastrous unpopularity".

Milk

Yields per dairy cow are increasing; so is the average size of dairy herd. But net profits of dairy producers are falling in real terms (that is if inflation and the relationship between farmers' costs and the market price for their milk products is taken into account). More milk is being turned into butter, but butter consumption has been falling even faster than milk consumption. Meanwhile, the cost of storing and disposing of milk takes up about forty percent of the total EEC farm budget. Between £90 and £100 is spent on subsidies for each dairy cow in the Common Market.

Just at the time when a proper understanding of agriculture in Britain is most needed, and is most lacking, the EEC's farm policy is giving agriculture a bad name. Stories of food mountains and wine lakes have dominated the headlines. The EEC dumps unwanted food on world markets disrupting the agriculture of other countries, notably those in the Third World. More than £3,000 millions are spent each year storing or disposing of EEC milk products which people do not want, or cannot afford to eat.

EEC farm prices are set at a high level in order to try and help the small producer as well as the large ones. But prices are common for all farmers, large and small. All farmers are

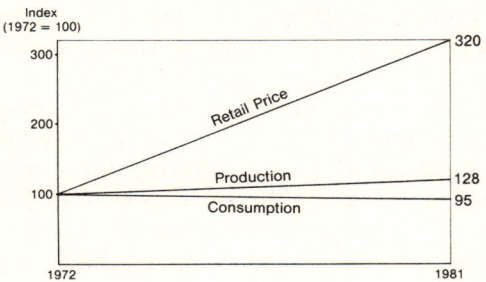

21. Milk Trends: *Milk consumption is falling, mainly because of high prices. In spite of this, milk production is increasing, and Britain remains a net importer of dairy products. As a result, more milk is used to produce butter, of which consumption is also falling, partly because of the controversy over health and diet. The cost of getting rid of the Common Market's milk surplus (eg by dumping products on Third World markets) takes up 40% of the EEC's agricultural support budget. Every cow in the Common Market costs almost £100 a head in subsidies. (Source: Milk Marketing Bd. & M.A.F.F.)*

protected in exactly the same way with the inevitable result that the gap between the rich and the poor increases. All products are guaranteed by high prices, including those in surplus. The system encourages more and more capital expenditure and widens the gap between the larger and the smaller producers. The bigger the farm, the more the farm produces, so the more money the farm is going to get out of EEC funds. And if prices are held down below the level of inflation or of rising costs, pressure increases on small farmers to produce more to maintain their standard of living.

There is increasing pressure on European farmers to cut down the production of 'surplus' products, including milk, fruit, vegetables and olive oil. Governments cannot afford, either politically or economically, to continue making available open-ended financial support. The smaller farmers are going to suffer on the grounds that they are 'inefficient' and cannot afford a cut in prices. The entry of Greece and Spain into the Common Market will compound the problem. Spanish landowners are already digging up olive groves, which have provided communities with a source of employment for thousands of years, to replace them with arable crops which need much less labour but whose high prices offer the prospect of a more lucrative future for the landowner in the EEC.

Apart from a few hand-outs for hill farmers and those in what are referred to in Brussels as 'less-favoured regions', and grants to 'modernise' (a euphemism for enlarging) farmholdings, price levels are the only instrument in the EEC's farm policy. And because in Britain, the main argument is over how much more British taxpayers have to pay into the Brussels budget, British Ministers are preoccupied during Common Market negotiations with securing easily quantifiable 'benefits'. These are most easily expressed in how much Britain will claw back from Brussels in financial terms, whether or not this is in the interests of British agriculture, let alone the economy or consumers as a whole. Peter Walker, the Conservative Agriculture Minister, in common with his Labour predecessor, John Silkin, returned from Common Market bargaining sessions waving a new EEC-financed butter subsidy for British consumers which will partly cushion the impact of price rises, even though the reason for the subsidy is primarily to get rid of part of the milk surplus, and in spite of the fact that butter consumption is falling and that the Department of Health, however cautiously, is recommending people to cut down on their fat intake.

Encouraged by high EEC prices, the Government enthusiastically backs the National Farmers' Union in its attempt to expand the production of sugarbeet in Britain: despite recommendations that we should consume less sugar, the fact that the climate and soil conditions are not suitable for sugarbeet, that the Common Market has a mountain of surplus sugar it dumps on the world market (even though the EEC at the same time gives grants for sugar factories in the Third World), and in spite of Britain's political commitment to import cane sugar from developing countries in the Commonwealth. Increased sugarbeet production would be an objective if the aim is as great a degree of self-sufficiency as possible. But the argument is not conducted in these terms; policy-makers do not take decisions with this aim in mind.

These two examples, of butter and sugar, illustrate what is at best the wholly superficial, at worst the cynical, attitude taken by Whitehall towards EEC agricultural policy. But there is also a great myth perpetuated about the EEC's farm policy: that is, it is founded on the principle that Europe must be self-sufficient in food. To begin with, it imposes high tariffs on such products as pineapples which do not grow in Europe. More significantly, given the predominant argument that high prices are worth paying in return for security of food supplies, the EEC is far from self-sufficient. Indeed, it is the world's largest food importer. The United States sells more food to the EEC than any other trading block in spite of Europe's protectionism. The Common Market imports about £3,000 millions worth of agricultural produce from the US each year, with Britain accounting for about £600 millions, roughly the same amount in value terms as the US exports to the Soviet Union.

The EEC imports three times as much food as it exports. The point is that a great deal of these imports — some 50 million tonnes a year — consists of animal feed. As a result of Europe's intensive livestock feeding systems, EEC farmers, including British farmers, are indirectly farming thousands of acres of the American mid-West and the millions of tonnes of maize and soyabeans grown there for Europe's animals.

The reason is that animal feed from outside the EEC (and this includes tapioca, or manioc, from Thailand which Thai farmers are busy growing for Europe rather than concentrating on the needs of the local population and of neighbouring countries) is much cheaper than animal feed grown inside the Common Market. The irony is that while European livestock farmers import feed, high EEC prices are encouraging European arable farmers to grow more and more high-yielding varieties of wheat which is only suitable for animal feed. Much of it is piled up in store, and eventually sold off at cost prices.

Grain, the most staple food, must be at the heart of any agricultural policy. The EEC started on completely the wrong foot. On the insistence of West Germany — not France, as is commonly supposed — Common Market governments agreed to set a high price for grain when the common farm policy was first established in 1966. As a result, arable producers have benefitted at the expense of livestock farmers, who feed their animals on grain, and who in any case generally have smaller holdings. Since Britain joined the Common Market in 1973 the acreage devoted to wheat has increased dramatically. By 1981, wheat accounted for nearly half the total cereals harvest. Over the past five years returns for wheat growers has risen by more than a half, significantly higher than those for other commodities. Wheat is a staple product; in some ways it represents the most 'efficient' use of the land — it needs less labour for instance. But present policies also encourages monoculture and specialisation and discourages mixed farming.

The high prices of cereal-based animal feed penalises the livestock farmer who is becoming farming's poor cousin to the rich cereal barons. Most of the wheat grown is of the high-yield varieties, most of which is not used in the standard British white loaf. Within three years of its commercial introduction in 1972, more than a third of the total wheat crop was sown with Maris Huntsman dwarf variety fit only for animal feed and processing for the food and biscuit industry. The price for wheat is attractive, says a senior Government scientist; "that is why the acreage is increasing, not because we need more of it". Of the 18 million tonnes of cerealsgrown in Britain, over two-thirds is devoted to animal feed, or stored for export. That proportion accounts for 2.4 million acres of good quality farmland. And while millions of pounds are spent each year on research into new, more productive, wheat varieties, very little is spent on the 90,000 other possible food plants. This massive switch to arable farming and appetite for higher yields has led also to the loss of hedgerows, old meadows and pastures.

Wheat production in Britain: (million tonnes)

1939	1944	average 1968-70	1977	1980
1.1	3.2	3.6	5.2	8.4

The Government's attempts to spell out its agricultural policy have been few and far between, and their contents pretty meagre. In 1975, it produced a White Paper entitled, "Food From Our Own Resources". The whole emphasis was on economic considerations and the supposed need for ever-higher yields. Nutrition was ignored. The document urged the expansion of "those commodities which seem capable of yielding the greatest economic returns to the nation". By implication, it also proposed a substantial increase in livestock production. Dr Pereira, then chief scientist at the Ministry of Agriculture, acknowledged a year later that if these official guidelines were kept, Britain would become more dependent than ever on imported feedstuffs. The annual 2½ percent increase in food production the government forecast proved far too optimistic and although the Government blamed it all on the 1976 drought, this was far from the only reason.

Four years later, in 1979, the Government produced another White Paper, "Farming and the Nation". This devoted a few token phrases about the need to protect the environment and a half-hearted reference to the Government's task to seek "to inform and persuade on the need for healthy eating". Otherwise, it appeared to accept as inevitable the move towards more and more highly-processed food, and laid great emphasis on the 'efficient, low cost' production of food in Britain compared to other EEC countries. Growth, it said, should come mainly through increased productivity, a euphemism for less labour and more machines and agrochemicals. It was a superficial and complacent 35-page document. Again assumptions proved ill-founded. An accompanying document — "Possible Patterns of Agricultural Production in the United Kingdom in 1983" — suggested that the acreage devoted to wheat, by 1983, would total 3,060,000 acres. The June 1980 agricultural returns for England alone showed that the wheat acreage had already increased to 3,379,200 acres.

Although the acreage is falling, grass still accounts for nearly 70 percent of Britain's agricultural land. (See fig 8, p.57) But even that is neglected. Production of meat and milk from British grassland could be doubled with more careful management. The poorer, peripheral, areas of the country are precisely those most dependent on grass. But rising costs and milk surpluses suggest that they will have to look to alternatives, including forestry and amenities, to attract tourism. Unless, that is, a radical re-think leads to the restoration of mixed farming.

Another crop which has found favour with farmers is oilseed rape, with the acreage increasing from 9,600 acres to 223,200 acres since Britain joined the EEC. It is a useful rotation crop and could save on imports of oil and protein for animal feed. But this, too, like wheat and sugarbeet, is the product not of a carefully-thought-out policy to meet the aim of greater self-sufficiency, but merely of high prices guaranteed by Brussels.

We have seen how, despite the rhetoric, the Common Market has no consistent policy towards greater self-sufficiency, either for Europe as a whole or its individual countries. For Britain, the situation is even more confused. British millers in the early 1980s continued to import 2 million tonnes (about £200 millions worth) of hard wheat from North America to produce the standard British white loaf. The wheat is imported even though it is expensive here since it faces high tariffs imposed by the Common Market. And not only does Britain maintain its commitment to import quantities of butter from New Zealand; EEC payments to encourage dairy farmers to get out of milk production so as to help get rid of Europe's massive dairy surplus have been eagerly sought by many British farmers.

British dairy producers, caught between high costs and consumer pressure to keep down prices, have seized the opportunity to switch to arable production which has the added advantage of being less labour-intensive. The dairy sector, with its system of monthly milk cheques, has traditionally been a main route into farming. It is now contracting in spite of the fact that Britain is a net importer of dairy products, relying on outside suppliers for about 30 percent of its consumption.

Food

Average household expenditure on food (excluding expenditure in restaurants and alcohol) as percentage of total consumer expenditure:

> 1938 : 27
> 1977 : 19
> 1979 : 20

Source: Ministry of Agriculture.

There has been a steady shift towards the consumption of more convenience foods, especially take-away foods. Consumption of chocolates and crisps is also increasing.

Before the war, only a third of all food bought in the shops was home-grown; the figure now stands at about 68 percent for a population that has increased during that time from 47 millions to 56 millions. But most statistics on British trade in food are presented in terms of value, not volume. The Ministry of Agriculture admits that, measured at constant prices, Britain's self-sufficiency ratio has not 'fallen' though this gives us no more comfort than the somewhat similar observation that if the relative price of oil had not risen, Britain would have had no balance of payments problem'. In volume terms, Britain's self-sufficiency ratio has increased only minimally in the past few years, and actually fell back in 1979.

While Britain produces more arable crops, notably wheat, its cattle herds are falling. Britain imports about £1,000 millions worth of meat and £500 millions of dairy products a year, and in terms of both value and volume, exports of food have increased faster than imports over the past decade. In 1978, exports of food, beverages and tobacco totalled £3,000 millions; imports of similar products amounted to about £6,000 millions, or 15 percent of all imports, the largest single sector of imports — even machinery as a whole came to £6,300 millions, only a little more than food and drink.

Two of the more significant figures show that between 1972 and 1977, the value of Britain's imports of fruit and vegetables increased from £429 millions to a staggering £1,108 millions, and of animal feeds £84 millions to £210 millions.

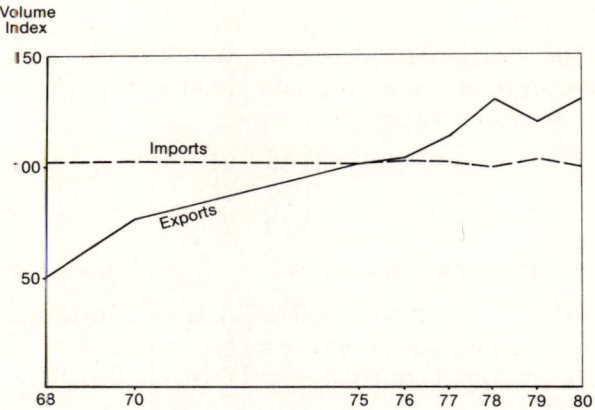

22. **Towards self-sufficiency?** *British farmers and fishermen produce just over half of the food consumed in this country, or about two-thirds of the types of food that can be grown in Britain. About £3,000 millions worth of food products are imported annually. The aim of self-reliance is accorded no more than lip service. (Source: M.A.F.F.)*

The British Government now is encouraging the food industry to export as much as possible — partly because, in this way, Britain's contribution to the EEC budget would be reduced for the industry would benefit from EEC export subsidies.

British policy-makers have paid little more than lip-service to the goal of greater self-reliance, as much self-sufficiency as possible, of a rational agricultural policy whereby farmers would produce a variety of crops which the population wanted, required, and needed.

Current priorities were made apparent by Peter Walker in what he described as an important policy speech at Wye Agricultural College in Kent on October 23, 1980, on the theme "agricultural and food prospects between now and the year 2000". He pointed to the increase in yields and productivity — expressed, of course, in terms of labour (per man employed) — that agriculture had achieved. He spoke of the technical advances in the field of embryo transfers for cattle, of the possibility of the speedy fattening of boars, of genetic engineering. He also mentioned the prospect of driverless tractors, but at the same time of the need to save energy.

He painted a picture of a Britain which, he insisted, farmers and the food industry would ignore at their peril. They must recognise, Walker stressed, "a major advance in the role of convenience foods suited to the single person household and the household where man and wife are at work; the micro-wave oven revolution . . . that will bring the need to adapt the packaging of processed meals to meet its needs"; the growth of deep-freezer ownership; "the ever-increasing expansion and diversity" of the fast-food industry "demanding new methods of the food supplier"; new canning and preservation techniques, the role of substitute foods and fish farming, especially for high-priced species. Walker seemed to see nothing incom-patable with this, and to what he referred to as "a major role" to be played by "health foods in all their forms".

The Government will therefore push agricultural policy towards this modern approach, urging farmers to be more competitive and market their produce aggressively, and food processors blindly to accept technological developments and to concentrate on valuable export markets. Such a shift of emphasis, officially encouraged, was amply reflected when Walker's first Royal Show in July 1979 was opened by Sir Hector Laing, chairman of United Biscuits. The time had come, Sir Hector said, when the food industry and farmers should get together and realise that their interests were identical. But he made it quite clear that the importance of the primary producer, the farmer, was declining in his view. He took the opportunity to advertise his company's latest initi-ative: a new line of quick-frozen convenience foods, including scampi-filled pancakes. The theme at the Ministry of Agri-culture's stand at the following year's Royal Show also reflected the new priorities: exports, with prominent displays of iris bulbs and the intensive cultivation of lobsters.

This approach and these priorities boost the interests of traders and food processors at the expense of the farmers. These interests are to encourage the development of new types of food and food processing techniques in order to maintain profit levels in a saturated market. "Added value", as Bird's Eye says, "is the magic ingredient". Thus the large companies produce more extravagant, quick-frozen meals, and textured vegetable protein, a euphemism for what is basically a can of soyabeans. Sir David Orr, chairman of Unilever, claimed in the course of the 1980 annual lecture to the British Nutrition Foundation (partly funded by the food industry) at the Royal College of Physicians in London: "the alternative to processed food is often not fresh food but no food at all; or at least a considerable restriction in the variety of food available". This is a preposterous exaggeration and begs a whole host of questions about policy, prices, distribution and supply — indeed, about the whole farming system. While research on ways to preserve food is clearly of benefit, processing food is a completely different matter. The number of lines in processed food increased from about 1500 in the 1950s to over 10,000 by the late 1970s. Yet the raw material for these products was almost exactly the same; the number of different lines, or food mixtures, is not synonymous with variety.

In parallel with this emphasis on processing and refining food, in the interests of profit rather than nutritional or agricultural needs, British policy has also encouraged the mass production of food. The aim is to produce as much as possible as quickly as possible at the lowest possible price. The Ministry of Agriculture and large companies proudly point, for instance, to the increase in crop yields. This has led to intense and increasing competition. (See fig 22, p.136)

The poultry industry offers one of the most dramatic examples. The consumption of broilers has increased from 25 millions a year in 1955 to about 400 millions in 1980. The consumption of turkeys over the same period increased from 3 millions a year to 23 millions. Those companies which produced most at the fastest rate, and which could afford expensive advertising campaigns, took over smaller companies and were applauded by the government for offering the twin prospects of cheap food and secure employment. Thus the Imperial Group now dominates the market in poultry. But in the space of three weeks in 1980 it announced the closure of three plants in Scotland, England and Wales. At the same time, Marshalls, the largest Scottish producer, closed a factory near Glasgow. Consumption had reached saturation point: 12 million chickens were piled up in cold stores up and down the country.

There was both a chicken and an egg glut. The Office of Fair Trading investigated the trading practices of Daylay Eggs, a subsidiary of the Imperial Group, during the egg price war a few months earlier. In the meantime, intensive livestock farming, notably in the poultry sector, is defended on the grounds that it is the fastest and cheapest way of producing the biggest amount of food.

Bread and Diet

This criterion — the fast, mass production of food — is also reflected, perhaps even more strikingly, in the story of the standard British loaf. The sliced white loaf has been promoted by Government in the interests of providing a relatively cheap and reliable staple food. The trouble was that the desire for uniformity leant itself to a mechanical process, while a stop-go policy on subsidies, the discount war in High Street super-markets and the cost of machinery killed off many smaller baking businesses. Small bakers were also hit by successive price increases of flour, of which the market is dominated by large corporations. Two of them, Ranks Hovis McDougall and Associated British Foods, are also large bakers. They mill flour for their own needs and offset losses from baking by the profits on flour milling.

The standard loaf is made on the instant dough principle[2]. From the start of dough-mixing to the end product is a process that takes about 90 minutes. A few minutes intensive dough-making is substituted for the traditional three-hours' fermentation. The priority for the large plant bakers is the biggest possible production in the fastest possible time. But then people stopped buying so much bread; in 1978, Spillers dropped out of the baking business with a loss of 8,000 jobs, while ABF and RHM was left in control of the market promising the government, which was worried about the monopoly aspects, to keep on some of Spillers' workforce for at least a year. The two companies kept the promise, but shed the workers shortly afterwards. In addition, the working conditions inside some of the large plant bakeries are not good the industry has had a bad safety record which has caused considerable concern to the Health and Safety Executive. It

Bread

Two large corporations, Associated British foods (ABF) and Ranks Hovis McDougall, dominate the British bread market. The other big miller, Spillers, now merged with Dalgety, withdrew from baking in 1977.

Three large corporations, ABD, RHM and Spillers/Dalgety, supply about 80 percent of the flour sold on the British market.

In 1977, the Monopolies Commission said: "We conclude that ABF, RHM and Spillers, by each requiring its flour-using subsidiaries to buy their requirements of flour from the group's own mills, are so conducting their respective affairs as to prevent or restrict competition".

ABF is part of a Canadian-based empire headed by Gary Weston whose father introduced sliced white bread in Britain, securing a new market for Canadian wheat. Spillers/Dalgety own large pig farms, securing a market for its grain. RHM, which has invested heavily overseas in food companies over the past few years, also makes pasta and noodles and other products new to Britain.

The large bread companies have always insisted that they produce white bread because that is what people want. The companies have ensured that it is *relatively* cheap. White bread is made from low extraction flour with all the wheatgerm, most of the fibre and some of the vitamins taken away (and sometimes sold separately, as in the case of wheatgerm, to the health foods industry). Some vitamins and additives are then added or put back in the course of a highly mechanised process which mixes the dough very fast, and reduces the fermentation time to a minimum.

Yet the traditional standard British white loaf, according to the latest available figures in the late 1970s, is 75 percent air by volume and 40 percent water by weight.

In 1980 while RHM spokesmen were still insisting that most people do not like wholemeal bread, it was spending £1½ million on an advertising campaign for wholemeal bread. That same year, the Royal College of Physicians recommended a high-fibre diet. Lack of fibre, it suggested, led to constipation, piles, and bowel diseases. Eighteen months later, the bread companies were selling white bread 'high in fibre'.

Total bread consumption (oz per person per week):

1955 : 55.1
1966 : 51.08
1978 : 32.13
1979 : 31.35

White bread consumption fell by about 20 percent during the 1970s, and consumption of brown bread increased by about 25 percent.

Sources: Ministry of Agriculture, including National Food surveys, and D. Hollingworth, 25 Years of Change, BNF Nutrition Bulletin, 4, 1978.

took the companies a great deal of persuading by the Executive in 1979 before they finally agreed to set a limit — of 115 degrees fahrenheit — to the temperature at which workers could enter the giant ovens in the event of a breakdown. Every minute out of action 'cost' hundreds of loaves in lost production, and therefore profits. The problems were compounded because of the all-night working shifts.

These moves towards the uniform standard loaf developed in the 1950s and was promoted in particular by Garfield Weston, a rich Canadian millowner who came to England and built up what is now the large ABF empire. It was given further impetus by the Chorleywood Baking Process, developed by the industry's research organisation in 1962.

This mechanical-chemical process saves time and money; it allows bakers to use weaker, and therefore cheaper, flour, with more water added. As one of the spokesmen of the independent master bakers put it: "the big plant bakers bake the bread quickly to keep the water in and the weight up". A white standard loaf in the late 1970s was 75 percent air by volume and 40 percent water by weight.

Chemicals are added for economic and commercial reasons. For instance, flour bleaches and improves naturally if stored. Most commercially-produced flour is bleached and improved artificially to save storage costs. Many of the chemicals used in bread are controversial; some have proved toxic in laboratory tests in animals, others have been banned outside Britain. Expert committees sponsored by the government have advised that some of the chemicals should no longer be used.

Chemicals used in the production of the standard white loaf include benzoyl peroxide, chlorine dioxide, and the anti-oxidant, butylated hydroxytoluene (BHT). Liquid paraffin, which can withstand a higher temperature than fats, is sometimes used to lubricate baking tins.

Most white flour is milled at an extraction rate of between 72 percent and 74 percent. The wheat germ and bran are removed and frequently sold separately either for animal food or for the health food industry. The vitamins and mineral nutrients for a 70 percent extraction rate are lost in the following percentages:

vitamin B_2	80 percent
niacin	80.8 percent
vitamin B_6	71.8 percent
magnesium	84.7 percent
potassium	77 percent
sodium	78.3 percent
iron	75.6 percent

Official Government standards mean that some of these vitamins and nutrients have to be put back in the course of the baking process. But a government White Paper issued shortly after the war noted: "the advantage of natural over reinforced foods is something about which most nutritionists in the United Kingdom are agreed. The high standard of health which has maintained in this country throughout the war, as already indicated, is believed to be attributable at least in part to this policy and to the consumption of high extraction flour in which the vitamins necessary for adequate nutrition have been retained".

Equally significantly, in the light of recent research into the importance of natural fibre in the diet, refined white flour gets rid of much of the original fibre in wheat. Respectable, albeit still controversial research, has attributed lack of fibre in the diet to the high incidence of diverticulosis, cancer of the colon, varicose veins, haemorrhoids and coronary heart disease. A report on the medical aspects of dietary fibre published in 1980 by the Royal College of Physicians recommended that people should be encouraged to eat food in an unprocessed, fibre-rich, form. It added: "a high-fibre diet in the sense of one based on unprocessed starch foods would also be low in fat, cholesterol, and animal protein and would be protective against large bowel cancer by all current theories".

A more telling point was made by Dr Ian Kennedy in his 1980 Reith Lectures. "The food industry", he said, "has been enormously successful in ruining our diet and, consequently, our health. For example, more and more food is made from raw materials which have been refined. Look at bread. The ordinary loaf of white bread is not lacking in nutrition. What it lacks is bulk. This lack encourages over-consumption. Over-consumption sells more bread and so produces greater profits. It also produces obesity and the consequent threat to health. And the story has a twist in the tail. The food industry has gone on to develop a second string to its bow: the sale of health foods and vitamin tablets. To make up for what has been removed from some food by its being processed and refined, vitamins and health foods, which are, of course, more expensive than ordinary food, then appear on the market. Thus, the wealthier can get the food they need, since they can afford to buy it. The poor cannot, and another inroad into health promotion and the prevention of illness is made".

The big bread companies say that they produce white bread because people like it. It is true that many more people buy standard loaves than brown or other types of loaves. But the white standard loaf is much cheaper. In addition, the consumption of bread has fallen by 2 percent a year over the past twenty years while over the past few years brown bread's share has increased from about 5 percent to about twenty percent. At a seminar on bread in 1980 Eric Towers, chief public relations executive for RHM said of wholemeal bread: "most people don't like it". Yet RHM had just launched a £1½ millions advertising campaign for its new range of brown breads. But there is yet another side to the story. There is widespread confusion about what different names of different brown breads actually mean. Some have wheatgerm added, some are coloured with caramel, the label 'wheatmeal' is meaningless. Peter Walker, the Agriculture Minister, has now recognised that consumers could be deceived.

Food companies rely on governments to continually emphasise, and pay respect to, the notion of 'freedom of choice'. As the Centre for Agricultural Strategy has put it[3]: "the individual is no more free to buy fat-reduced milk, bread without added calcium, foods with prohibited additives, home-produced butter without colourants, or beer without paying one-third of the price as tax, than he or she is to harbour rabid dogs, or sell narcotics".

In spite of their protests that they do not interfere with the people's freedom to choose, governments and companies, of course, do so through the whole range of taxes, subsidies and the allocation of resources at their command. What is lacking in Britain, at least among the policy-makers, is a consistent attitude towards good nutrition and food production, diet and agriculture, health and taxation. The government does not impose high taxes on tobacco, or control tobacco advertising more strictly, because it does not want to provoke such a sharp drop in smoking that would reduce the revenue it gets from tobacco sales. Equally, Whitehall officials have advised Ministers that a decline in smoking would increase the number of elderly people likely to 'impose' extra burdens on the health and social security services. A similar attitude is adopted towards alcohol, which is now much cheaper relative to both average wages and to other basic foodstuffs — notably bread — than it was twenty years ago.

Food and Health

The lack of knowledge about food and nutrition appears to be deliberately perpetuated by the Government. The Department of Health and Social Security says that its role is to provide information to enable consumers to make wise choices about diet, but the information available is conflicting and inadequate. While the DHSS suggests that we cut down on fat intake, the Ministry of Agriculture's food standards committee publishes a report on labelling which says that food manufacturers should not be obliged to show how much fat their products contain. According to one senior member of the committee, Dr Philip James of the Dunn Clinical Nutrition Centre at Cambridge, Britain's food labelling system is "antiquated and even mediaeval".

The consumer has little knowledge of food additives and little chance to avoid them, even though about half of the 3,000 or so additives used in Britain are merely cosmetic. The DHSS list of recommended intakes for nutrients is virtually useless, either as a basis for analysis or as an instrument to measure the diet of different groups. It is far too general and does not describe which nutrients are found in different foods. The Government admits in its National Food Surveys that the quality of the national diet and nutritional standards are falling. Yet it says that there is no cause for alarm. However, independent nutritionists and scientists say that it is impossible to know how the government arrived at this conclusion, particularly since the nutrient intake per head in households with children is consistently below the recommendations for some nutrients. The Centre for Agricultural Strategy has said [4] that a rising proportion of the population is at risk but "at present there is no published basis for determining the nature of the increasing risks or the appropriate degree, or form of concern".

Whitehall, backed by both food and drug companies, concentrates on trying to cure symptoms but neglects the need to prevent diseases arising out of malnutrition, including obesity. Britain is far behind other industrialised countries when it comes to trying to prevent diet-related diseases, as well as heart diseases, many of which are now recognised to be linked to diet, one of the commonest causes of death in men over 35. In West Germany, 2 percent of the Gross National Product — about £1,000 millions — is spent each year helping people to recover from over-eating and over-drinking. The figure for East Germany is £500 millions. In Britain, according to the British Nutrition Foundation, which may underestimate the problem, diet-related diseases cost the National Health Service £550 millions a year.

There is little effective cooperation between the DHSS and the Ministry of Agriculture, two important policy-making departments. The Government, as we have seen above, negotiates an EEC-financed subsidy on butter because of political considerations, an attempt to limit price increases and the need to get rid of surplus produce. Consumers have to buy milk with a high cream content: an example of unnecessary, indeed harmful, controls over the quality of food. On the Continent, there is a wide range of liquid milk available, including skimmed milk.

A more encouraging sign is that butchers and consumers are beginning to persuade farmers and the government to reduce the fat content of carcasses, a move that will also save producers a lot of money in feed costs. It is an example of how consumers can use their influence over farming practices.

All this must be set against the background of advice from at least 18 different committees, ranging from the New Zealand Heart Foundation to Whitehall's Committee on the Medical Aspects of Food Policy (known as 'Coma'), that we should eat less fat and consume less sugar. But marshalled against these are our old enemies, the food companies. To begin with, they have enormous advertising budgets which Alistair Mackie described shortly before he was advised to resign from his post of director-general of the Health Education Council in 1980 as making a mockery of Britain's advertising standards system. Three companies — Rowntree Mackintosh, Cadbury and Mars in 1978 spent £34 millions on advertising. Against this, the Health Education Council's total budget amounted to just £4.75 millions, and while the companies increased their expenditure, the Council's budget was cut by a quarter.

Six food companies in 1980 were allocating £6 millions between them to promote noodles produced with a mixture of wheat and starch. The British Poultry Federation launched a £600,000 television advertising campaign, the Eggs Authority at the same time earmarked £1.75 millions to boost sales. RHM and ABF spent more than £1 millions trying to arrest the long-term decline in consumption of white bread before allocating £1½ millions to promote brown loaves.

Top advertisers: (source: *Campaign,* 27 July 1979)

1	Rowntree Mackintosh	£12.1 millions
2	Cadbury	£11.86 m
3	Mars	£10 m
8	Brooke Bond Oxo	£6.77 m
13	Van Den Berghs (Unilever)	£5.84 m
14	Pedigree Pet Foods (Mars)	£5.77 m
23	Kellogg	£4.6 m
24	General Foods	£4.5 m
27	National Dairy Council	£4.2 m
31	Bird's Eye Foods (Unilever)	£3.9 m

Companies invariably promote an image of fresh national foods, with suitable pictures of the countryside, in their advertising. Just as other entrepreneurs have invested heavily in the tourist and 'leisure' industries, so food companies have invested in health foods and 'hot bread' shops attached to the supermarkets. They have set up institutions with impressive titles such as the Flour Advisory Bureau, sponsored by the large bread companies, and The Butter Information Council,

The Food Leaders

The world's biggest food and beverage processing companies in order of size. Figures relate to 1976 — the most up-to-date available.

PARENT COMPANY	HOME COUNTRY	FOOD PROCESSING REVENUE $US mill.	TOTAL REVENUE		NET INCOME	
			AMOUNT - $US mill.	PROPORTION FOREIGN - %	AMOUNT - $US mill.	PROPORTION FOREIGN - %
1 Unilever Ltd	Nld/UK	7,900	14,800	71	1,276	51
2 Nestlé	Switz.	6,248	7,248	95	–	–
3 Kraft	USA	4,776	4,977	15	136	2
4 General Foods	USA	4,402	4,910	26	177	15
5 Esmark Inc (Swift)	USA	3,955	5,301	16	83	14
6 Beatrice Foods	USA	3,943	5,289	21	183	20
7 Coca-Cola	USA	2,911	3,033	44	285	55
8 Greyhound Corp	USA	2,385	3,738	–	77	–
9 Ralston Purina Co	USA	2,365	3,394	24	126	14
10 Borden Inc	USA	2,336	3,381	16	260	20
11 United Brands Co	USA	2,130	2,276	26	16	–
12 Iowa Beef Processors	USA	2,077	2,077	0	27	–
13 Imperial Group Ltd	UK	2,070	5,790	12	132	–
14 Archer-Daniels-Midland	USA	2,065	2,118	27	61	–
15 Pepsico	USA	2,051	2,728	21	136	15
16 Assoc. British Foods	UK	2,050	3,012	–	–	–
17 Carnation Co	USA	2,004	2,167	15	104	19
18 CPC International	USA	1,968	2,696	55	122	55
19 LTV Corp	USA	1,919	4,497	–	31	–
20 Heinz	USA	1,882	1,882	41	74	34
21 Seagram	CAN	1,874	2,049	94	79	–
22 Rank Hovis McDougall	UK	1,801	1,861	13	95	20
23 Procter & Gamble	USA	1,800	7,349	25	461	18
24 Nabisco	USA	1,780	2,027	29	77	11
25 General Mills	USA	1,735	2,909	16	117	10

● Unilever, the world's biggest food business, has 25 per cent of its employees in Africa. In 1978 the company's turnover of $10 billion was greater than the combined GNP of 25 African countries.

From : *New Internationalist,* February 1982.

sponsored by the dairy industry and creameries. Food industries sponsor the British Nutrition Foundation. (It has referred, for instance, to the possible dangers of omitting breakfast citing a Kellogg breakfast survey of 1977 which also showed that 40 percent of those questioned ate a 'cereal-type' breakfast: clearly not enough).

The British Sugar Bureau together with the biscuit, cake, chocolate and confectionery industry federations contribute funds to the British Dental Association for research into 'alternative' causes of dental decay. They are spending thousands of pounds a year on research by the Royal College of Surgeons into a drug that would counter the risks of dental decay. Teeth decay costs the National Health Service more than £100 millions a year, and the consumption of sugar in Britain is one of the largest in the world. The DHSS pamphlet, 'Eating for Health', itself admits: "The United Kingdom has a poor record of dental health. It has never been good, but has become worse over the past century principally through the increasing use of sugar in the diet".

Sugar is a bulking agent, is addictive, and is an anti-coagulant providing colouring and preservative in jams, pickles and confectionery: it is therefore very useful to the food industry generally. Yet it has no nutritional value. John Yudkin in *Pure, White and Deadly*, says: "there are two undeniable facts about sugar: 1. there is no physiological requirement for sugar. 2. If only a small fraction of what is already known about sugar were to be revealed in relation to any other material used as a food additive, that material would be promptly banned".

According to the Health Education Council, "the assumption that sugar is a highly desirable food is taken completely for granted. Not only is it never questioned, but it is thoroughly reinforced". For instance, a study kit for parents, teachers and schools includes the assertion that "sugar is an important part of a balanced diet". A children's hand-out, called *My Tooth Diary* produced by the General Dental Council in conjunction with the Mars Health Education Fund (sic) refers only to the dangers of 'food' left on the tooth, without giving a specific warning against sugar. The Health Education Council suggests that warnings should be given, along the lines of: "this breakfast cereal contains 50 percent sugar and should be regarded as confectionery rather than food".

How has our diet changed since the war? The British are eating less butter, red meat, white bread and fresh fish. Consumption of poultry, brown bread and margarine is increasing. While direct consumption of sugar is falling, we are consuming more sugar in soft drinks, chocolate and sweets. Although our consumption of animal fat has dropped, we still obtain more calories from animal sources than from vegetables and our total calories intake is still substantially higher than recommended DHSS levels. The British eat less cereals and meat than people in most other West European countries, and much less fruit and vegetables. We eat more potatoes, and sugar, and drink more milk. Our diet is less carbohydrate-based, and far more convenience-food oriented. Our calorie levels are far in excess of our energy needs.

While we continue to spend more on food than on any other single item, the proportion of the average household budget spent on food has dropped from 27 percent in 1938 to about 19 percent today. There is widespread evidence that when incomes are hit, households cut back on food first, a tendency constantly referred to by farmers who, in turn, constantly point out that the real cost of food has fallen. In 1966, for instance, the average man had to work 14 hours to pay his household food bill; ten years later he had to work ten hours. While we are spending less on food in relation to other goods we are also eating less basic, staple produce. The way we are going and the way we are imitating the Americans, as much as 80 percent of our diet will shortly consist of convenience food. (And in the United States, one out of every three meals are eaten outside the home; ten years ago, the figure was one out of four.)

The life expectancy of the male population in the US actually fell during the 1960s and 1970s and a federal survey concluded: "a significant proportion of the population was malnourished or was at a high risk of developing nutritional problems". Large numbers of Americans were found to suffer from mineral and vitamin deficiencies and 40 percent were 'obese'. A US Senate committee chaired by George McGovern concluded in 1977 that a number of ailments, including heart disease, cancer, hypertension and tooth decay were linked to the American diet and could therefore be prevented. Fat consumption, particularly of saturated fat, and the consumption of sugar, it said, should be cut drastically.

Some new food products :

Products introduced 1945-65:

Vitamin-enriched breakfast cereals,
Dehydrated potato specialities,
Boil-in-bag frozen vegetables,
Liquid diet foods,
Freeze-dried soluble coffee.

Products introduced in the late 1970s:

Cough candy lollipops,
Yogurt bran bread,
Kosher bubble gum balls,
Spoonable cheese spread,
Powdered Worcester sauce,
Smoke-flavoured salt,
Frozen quiche.

Source : The Organisation for Economic Cooperation and Development (OECD) *Observer,* September 1980, which added that less than ten percent of the food sold by American supermarkets "had first been bought (or produced) and processed by a food firm. Although more highly developed in the United States, the trend for food processing to dominate agriculture is clear elsewhere as well".

It added : "The vast majority of the thousands of items marketed as 'new' owed their novelty only to a new brand name, changes in packaging or flavour, or other minor variations........ In addition, vast amounts of advertising are required in such a system of competition through product proliferation. In both the United States and the United Kingdom, food firms rank among the top advertisers."

Following an extensive debate among the public and in Parliament, the Norwegian Government has used taxes and subsidies for farmers specifically to influence people to adopt a diet which was widely accepted to be the best one for the health of the population and for the best standards of nutrition. Farmers are encouraged to feed their cattle on as much grass as possible, rather than cereals, to reduce the fat content. In Sweden, the nutrition department of the National Medical Society has drawn up a 'diet and exercise' programme whose recommendations includes a 100 percent increase in the consumption of green vegetables and a 25 percent cut in fats, oils, sugar, syrups and sweets. The Swedes recognised the importance of consulting agricultural interests, the food industry, teachers, health workers and the media. Cooperation produced a serious and calm debate, and in marked contrast to the British and the American industries (which strongly attacked the targets proposed by the McGovern committee) the Swedish margarine and dairy groups jointly financed and wrote a paper directed at consumers advising them to reduce their fat consumption. Low fat milk and cheeses are being produced, and leaner sausages and meat appeared in the shops.

In Britain, a 1978 CAS report, "Food, Health and Farming", suggested that the fat content of the British diet should be cut from 40 percent to 30 percent; that the fat content of milk should be halved; that financial incentives should be offered to farmers to produce leaner carcasses; and that there should be a maximum content of fat in processed foods such as sausage. Health considerations, it said, need not jeopardise farming interests.

But the prevailing official view in Britain is that the Government should limit itself merely to collecting and disseminating general information and let individuals form their own view and take their own decisions. Now we have seen that 'freedom of choice' can be virtually meaningless in this context. And while the Government officially takes a back-stairs, passive, role the food sectors confuse the issue further by arguing amongst themselves. For instance, the Butter Information Council says that while the consumption of butter has fallen significantly in the US over the past decade, the consumption of margarine has increased, and so has the incidence of death by heart disease. The weakness of labelling regulations, it adds, disguises the high content, frequently as much as a third, of animal fat in margarine. The British Nutrition Foundation — which is a registered charity, though, as we have seen, is sponsored by the food industry — argues that naturally occurring compounds in food are more likely to cause toxicity than intentional food additives. The food industry says that natural toxins can lead to apparently inexplicable allergies.

Experience in other countries, including Sweden, where 'nutrition' is not such a dirty word, suggests that it takes a long time to encourage people to be aware of sensible eating, a more rational diet, and to be sceptical of the claims and suspicious of the priorities of the food industry. The industry, after all, is motivated by the search for profits; this is also the motivation behind the company's research programmes. But in Britain the problems are compounded by the attitude of Whitehall whose passion for secrecy, natural bureaucratic hostility to sharing information with the outside world, and obsequious attitude towards large companies, extends to food and agricultural policy as it does to most other issues.

The Food Standards Committee, the Food Additives and Contaminants Committee, the Committee on the Medical Aspects of Food Policy — all these groups of scientists, representatives of the food industry and civil servants — meet in secret. Many academics involved in food research are themselves sponsored, indirectly or directly, by large food companies. In deference to the view that commercial confidence must be protected, Whitehall has decided to bypass existing committees and set up secret machinery to vet novel foods, based on nutrients from grass, for instance, vegetable residues, and even manure. Awards of certificates to allow companies to make new products will be secret both to the general public and to Parliament. Recommendations may be published in general terms, but the justification for them are not disclosed.

We can now turn to related problems which reflect these priorities, problems of pests, pollution and waste.

6

PESTS, POLLUTION AND WASTE

In November 1980, a cloud of poisonous sulphuric acid escaped from the Royal Ordnance Factory at Puriton, near Bridgwater in Somerset. The factory is one of the main production plants for the manufacture of high explosives. Villagers at Woolavington, a quarter of a mile away, were vomiting, children were coughing when they came out to play at the village school. The Ministry of Defence said the leak caused only "minor irritation". A month later, after another leak of the acid, and with no warning to people living nearby, the Ministry said: "It is not worth getting excited about. It was all over soon after it started".

Earlier in the year, seven thousand people in Barking, East London were evacuated after three explosions at the Robert Wormsley chemical factory released highly toxic sodium cyanide fumes. The fire brigade did not know on that occasion, and in many other instances, what explosives and chemicals are stored or manufactured in factories, many of which are situated in densely-populated areas. Factory inspectors privately admit that many companies, both large and small, pay scant attention to health and safety regulations. They are bound by the Official Secrets Act and actively discouraged from disclosing what they see, by both ministers and senior officials at the Health and Safety Executive. Meanwhile, Whitehall says that cooperation between the Alkali Inspectorate and industry depends on secrecy. The Inspectorate is forbidden to give details of emissions from individual works on the basis that competitors must be prevented from discovering industrial secrets. The Inspectorate argues that, in any case, ordinary members of the public would not know how to interpret information even if they were

provided with it. The Alkali Inspectorate has withheld information on fluoride poisoning to the extent that scientists are
unable to assess past damage and comment on existing
dangers (notably to cattle).

There is a uniquely British mixture of complacency and
secrecy. We are far behind other developed countries in our
attitude towards working conditions, public hygiene and
safety. Perhaps many stories about dangerous incidents or
potential hazards from the nuclear industry are exaggerated;
but any inclination to give the benefit of the doubt is easily
swept aside by the arrogant and presumptious reaction of the
authorities and the 'experts'. They feel 'hard done by', they say
they have an 'unfair Press', from people who do not fully
understand what, after all, are complex technological matters.
The attitude is the same in the agrochemical industry.

There is no shortage of examples. In 1980 an official report
suggested that people living near coal-fired power stations may
receive a substantial dose of radiation from ash from the
chimneys. (The Ministry of Agriculture had not made any
measurements of food near the power stations because the
Radioactive Substances Act does not stretch to fly-ash from
coal.)

The London Brick Company in 1980 wanted to build new
works near Bedford, though a Ministry of Agriculture report
six years earlier showed that cattle in 19 out of 43 farms near
existing brick factories had fluorosis. (At Peterborough, 20 out
of 38 farms near another LBC works had the disease.) The LBC
is the largest member of the National Farmers' Union in the
Bedford area.

In 1979, the Department of the Environment warned the
residents of Shipham, a small village in Somerset, not to eat
locally-grown vegetables, and advised heavy smokers to cut
down on their tobacco consumption. The warnings were
sparked off by a survey by Imperial College, London, which
found very high concentrations of cadmium — a toxic
chemical that causes long-term kidney damage — in the soil:
the remains, apparently, of an old zinc mine.

The Department knew about the contamination six months
earlier. Surely, it was asked, it should have announced the
dangers then if they were so great. No; there would have been

similar delays even if the threat was even more serious. Time was needed to organise health diets and soil surveys. The Department provided no figures on the levels of cadmium, and made it clear that it had not looked at other villages close to zinc mines.

The established British approach towards pollution, both agricultural and industrial, is pragmatic. That is to say, avoid setting fixed and legal standards as much as possible; take every case on its own merits; do not follow practices in other countries; do not legislate what is apparently 'unenforceable'. Emphasise what is 'reasonable', and take into account the cost of pollution control and pay due deference to threats from manufacturers that stricter controls would lead to more unemployment because of the costs involved.

Death is the only proof of danger. The successful Clean Air Acts followed the death of 4,000 people during the London smog of 1952. Seventy former workers at the Cape Asbestos mill at Hebden Bridge died before the dangers of asbestos were recognised officially. The report on the incident by the Ombudsman noted that "no significant advance was made (at the time) in scientific knowledge of what level of concentration could be accepted as safe". In 1979 more than 2,000 sea birds were found dead on feeding grounds in the Mersey estuary. Analysis at the Institute of Terrestrial Ecology found organic lead in the birds. The water authority and Associated Octel, a company manufacturing lead additives for petrol at Ellesmere Port, said that they had found no traces of lead. A few months later, the company and the water authority acknowledged that higher than normal levels of lead had, after all, been found in the birds and in the birds' food.

An official report, "Lead and Health", was published in 1980, drawn up by a committee chaired by Professor Patrick Lawther of the environmental medicine faculty at St Bartholomew's Hospital Medical College. It said that the dangers of lead from food was much .greater than air-borne lead, 90 percent of which comes from petrol emissions. But this, the report said, accounted for only 10 percent of lead in people's bodies. Lead from paint and pipes was much more significant. It gave no firm recommendations, saying merely that lead in petrol should be "progressively reduced". Yet many other studies had pointed to lead in petrol and in the air as the cause of lead in food.

A little later, a study "Lead or Health", published by the Conservation Society, produced evidence, which, it said, the Lawther committee had ignored, that low-level lead pollution caused damage to children in inner-city areas, notably by inhibiting development of the nervous sytem. It said that by assuming that the amount of lead in someone's body was already high because of food intake, blaming this partly on natural lead in the soil, the official working party had artificially produced a much smaller rise in blood-lead lead levels from breathing than what in fact occurs.

Meanwhile, scientists on a Department of the Environment-sponsored report on lead in blood of people living near Birmingham's 'spaghetti junction' concluded that though the levels were high, they were not unusual for city dwellers, and there were no dangers. Yet scientists at the Harwell nuclear research establishment had stated back in 1978 that lead level risks from exhaust fumes could be twice as high as previously assumed and could effect the mental development of children.

Confidential reports compiled for the Greater London Council showed that blood-levels are significantly higher than the internationally-agreed safe minimum. The Lawther group did not take evidence from scientists in the US where concern about petrol fumes has already encouraged 40 percent of all motorists to use lead-free petrol (which costs just 5 cents a gallon more than ordinary petrol). The US, Japan, Sweden, the Soviet Union and West Germany have reduced the lead content of petrol. All cars sold in Australia after 1985 will run on lead-free petrol. Car manufacturers say that engines capable of running on lead-free petrol would put up the cost of a car by £500 and more crude oil would be needed if there was a switch to lead-free petrol. Yet both British Leyland and Ford have produced such cars for export to the US and Japan. The British Government argues that if it followed West Germany's example, it would cost the economy £200 millions a year, a fraction of our GNP. Professor Bryce-Smith of Reading University and a leading expert on chemicals and the environment says: "the human brain is the most complex organism in the universe, yet we allow this to happen. It could be stopped for an extra penny a gallon, the cost of lead-free petrol".

Further evidence about the dangers of lead poisoning came in thick and fast. A Newcastle survey demonstrated that there was a relationship between mothers' exposure to lead and abnormalities in their children. Dr Robin Russell Jones of St John's Hospital, London, disclosed that 95 percent of children in Britain had blood-lead levels higher than the figure at which harmful effects had been demonstrated by studies in the US and Britain. The recommended safe normal range was seven times higher than that considered now to be the safety threshold.

Estimates by the car companies of the costs in the US of a shift to lead-free petrol was found to have been exaggerated fivefold and in West Germany the cost to oil refineries for a shift in policy there was a third of the original estimates.

Faced with this battery of evidence, the Government in Britain announced in May 1981 that it had agreed to reduce the maximum level of lead allowed in petrol from 0.40 to 0.15 grams per litre by 1985.

The decision might have been more radical, and the Government would have been severely embarrassed, if a private memo written by Sir Henry Yellowlees, chief medical officer at the Department of Health and Social Security, had been made public. Written in March 1981, it flatly contradicted the Lawther report's contention that there was no link between lead in petrol and lead in food. "Lead in petrol", Sir Henry told Sir James Hamilton, permanent secretary at the Department of Education and Science, "is a major contributor to blood lead acting through the food chain as well as by inhalation".

"There is no doubt", he continued, "that the simplest and quickest way of reducing general population exposure to lead is by reducing sharply or by entirely eliminating lead in petrol". Hundreds of thousands of children were affected by the risk. Sir Henry referred to evidence of EEC research that indicated petrol lead may contribute on average about 27 percent of total blood lead in adults and about 40 percent of total blood lead in children. In spite of this, the Government was still telling MPs in Parliament that the figures were much lower.

Sir Henry went on in his letter: "There is a strong likelihood that lead in petrol is permanently reducing the IQ of many of our children". He confirmed what many people were assuming: "the Environment Departments, Health Departments and Ministry of Transport are recommending a very considerable reduction of lead in petrol, but this is opposed by the Department of Energy and the Treasury on economic grounds".

The British approach towards toxic chemicals, including agricultural pesticides, is based on the assumption that there is a 'safe' level of risk beyond which it is unacceptable, and below which the risk is acceptable. Experience — of DDT, for instance, or radiation from H-bomb tests in the atmosphere — has demonstrated that there is no such threshold. "Pesticides are by design biologically active, and hence hazardous chemicals: however stringent the tests applied to them, there is the possibility of unforeseen and unforeseeable effects", the Royal Commission on Environmental Pollution has pointed out[1].

This is even more likely when so many different pesticides are mixed or used separately on the land, itself a constantly changing, biological substance. Many pesticides are based on highly toxic substances, either organophosphorous chemicals — discovered by Germany during the war and a basis for nerve gas — or organochlorines, including aldrin and dieldrin which, though banned from most uses in the US, are still used in Britain. They are extremely persistent: not easily broken down.

The alternative view that there is in fact no threshold of acceptibility and that any exposure to toxic chemicals must be regarded as harmful to some degree[2] seems to be the only sensible one to adopt — even more so since scientists frequently admit that it is extremely difficult, even impossible, to establish complete information about the effects or the inter-action of different chemicals on crops or the soil. The Ministry of Agriculture's Advisory Committee on Pesticides itself admits that "there can never be total proof of the safety of any product of any kind . . . (and) dosage alone determines poisoning".

Chemical manufacturers, claiming the need for commercial

protection against their competitors, publish few details about their products. Much more information about identical chemicals are made available in the US. Neither manufacturers nor Government departments are willing to give detailed figures about the quantities of pesticides sold and used in Britain, with the result that research into their effects is made even more difficult. The reluctance of pesticide manufacturers to release information, Sir Hans Kornberg, chairman of the Royal Commission on Environmental Pollution and an eminent biochemist says: "reflects the quite unnecessary secrecy and confidentiality which seems to permeate so many aspects of life". In its report on agriculture, the Commission, which said that Britain was behind the US and Europe in its attitude towards both safety and efficacy of chemicals, added: "the data we have obtained to illustrate growth in the production and use of pesticides in either financial or tonnage terms are . . . without as much meaning as we would have wished".

Shortly after a country doctor became concerned when a number of patients who worked on the same farm came to his surgery complaining of the same symptoms, the *British Medical Journal* reported "four out of five members of a team of farmworkers who had been using various herbicides and pesticides in intensive agriculture became impotent. Sexual function recovered after further contact with the chemicals was stopped and hormone therapy had been given, though in one case this took about a year. We have not been able to incriminate one particular substance, but with circumstantial evidence and lack of any other obvious cause it seems likely that the impotence was due to the toxic effects of one or more of the chemicals being used. "The case has still not been solved. It seems that the employer followed Ministry of Agriculture regulations, but that the workers occasionally disregarded them because of the lack of ventilation in the protective clothing".

One of the most disquieting controversies involves 245-T, a herbicide that can contain the impurity dioxin, acknowledged even by the British Government to be one of the deadliest chemical compounds ever made. The way the controversy has been handled by the authorities is as significant as the product itself.

The Ministry of Agriculture has reviewed it no less than eight times and passed it safe so long as it is used "for recommended purposes" and that "recommended precautions" are taken. Yet its use for most purposes has been suspended in the United States following a survey of miscarriages in Oregon in 1979. The survey revealed an abnormally high rate of miscarriages among women living in areas which had been sprayed by the product. (Dioxin has also been found in fish eggs in the Great Lakes Basin.) In later summer 1979, Mrs Gillian Scheltinga miscarried a month after the Forestry Commission used 245-T near her home in the Gwydyr Forest, Llanwyst in Gwynedd. She had eaten blackberries from bushes sprayed by the Commission and described by its officials as being safe. Later she found that drums still containing quantities of 245-T had been dumped down a nearby mineshaft. The wife of the warden of an outdoor pursuits centre close to the mineshaft also suffered a miscarriage a short time before. The National Union of Agricultural and Allied Workers drew up a dossier of several cases offering some evidence of a link between 245-T and miscarriages, birth deformities and skin diseases.

In December 1980 Whitehall's Advisory Committee on Pesticides dismissed the results of some 20 reports, most of them conducted in other countries, which suggested that there is a link between dioxin and 245-T with cancer and birth deformities. The US Food and Drugs Administration for example had conducted laboratory tests with animals which showed that 245-T without dioxin can produce dead and deformed offspring. The Whitehall committee concluded that there was no valid or sound medical and scientific reason why, in its view, 245-T should not be used. Yet in the same year, 1980, it reduced the official 'safe' maximum of dioxin from 1 part per 10 million to 1 part per 100 million. The Ministry of Agriculture acknowledged that facilities for detecting the presence of dioxin are expensive and are not available in its own laboratories. Just a year earlier the Ministry's advisory committee acknowledged that it is "not satisfied that a reliable analytical method yet exists for the determination of such very low levels of TCDD (dioxin) in formulated products, given that these contain a number of other substances which interfere to a considerable extent with some analyses. (It) therefore recommends to Government departments that experimental work be carried out to provide a method for the determination of TCDD in formulations and thus to allow it to set a workable standard". Professor Robert Kilpatrick, the committee's chairman and dean of medicine at Leicester University, acknowledges that the best possible tests would need an epidemiological survey, but that would be very expensive.

It is not in anyone's interest to promote 'scare-stories' based primarily on emotion and pressure groups which may have one-sided motives. But the controversy over 245-T brought two broader issues out into the open. First, the Whitehall-sponsored group of experts admits that instructions about how the herbicide should be used, and what protection those who use it should have, are not always followed. Products based on 245-T can be obtained freely over the counter, often with warnings on the labels which are no different to those on a bottle of aspirin. In addition, the trade unions, notably the farmworkers', persistently point out that conditions in the field — with the possibility of spray drift, for example — are very different from those in laboratories where tests are taken and analyses made.

Secondly, the controversy illustrates the growing tension between 'experts' and 'amateurs'. John Home Robertson, Labour MP for Berwick and East Lothian, put it succinctly during a Commons debate on the herbicide in December 1979. The Ministry of Agriculture's committee, he said, consisted of ten academics and fifteen civil servants, a mixture 'of boffins and bureaucrats'. The unions want agricultural chemicals to come under the responsibility of the Health and Safety Executive, where unions are represented and where the use of other toxic chemicals is debated, and they want the Ministry of Agriculture's Advisory Committee on Pesticides to be abolished. For a committee that confidently prides itself on its objectivity and independence, its attacks on the unions for questioning its role — echoed in similarly sharp attacks by the British Agrochemicals Association — are depressing. Was this another example, as in the controversy over lead in petrol, of too many scientists becoming embraced by the industrial, economic and bureaucratic establishment? Throughout the debate over 245-T in 1980, both ministers and civil servants in the Ministry of Agriculture made it absolutely clear that for them the crucial issue was the threat to Whitehall's maze of expert, and secretive, committees and public confidence in them. It was this, as well as the substance of the specific controversy — the dangers to health and the land — which was vital so far as the Government was concerned.

As the *New Scientist* remarked on the subject of lead in petrol:

"Increasingly, politicians must be brave. They must learn that science does not have the answers to many questions involving risk. That is a difficult task. It will be interesting to see whether the report of the Royal Society's working party (set up in an attempt to tackle the problem of risks in a democratic society) has the courage to help politicians to make the leap or whether the working party will retreat behind the crumbling wall of scientific pretension".

It was significant, too, that when Professor Kilpatrick was asked why 245-T was banned in the Netherlands and Italy, had been temporarily taken off the market for most uses in the US, and voluntarily withdrawn from the market by producers in Denmark, he dismissed these initiatives as having been dictated by public concern and emotion. (In Britain, the product has been banned by 70 local authorities as well as by British Rail.) The Ministry of Agriculture's committee rejected the use of alternatives, such as glysophate, on the grounds that it would be more expensive. And the committee members were aware that employers are relying increasingly on chemicals as a means of reducing labour costs and a way of cutting down on manpower.

An irony in the 245-T controversy is that one of its specific uses is against nettles, a plant that actually feeds the soil with nutrients, and is a basic food for butterfly larvae.

Another controversy has broken out on both sides of the Atlantic over 24-D, the most widely used phenoxy herbicide and the sister chemical compound to 245-T: together they helped to make up 'agent orange', the defoliant used by US forces in Vietnam. Several hundred Vietnam veterans have filed suits, claiming injury, including deformities in their children, against the US Government, following alleged contact with the product (it is worth notice here that one of the nightmares the Whitehall bureaucracy — indeed, any bureaucracy — suffers is claims and lawsuits from members of the public). The US Health, Education and Welfare Department had concluded, after tests on 79,000 animals that 24-D produced as many deformities as 245-T. Both are plant hormones, which force plants to grow extremely quickly, and then die. Spraying of 24-D has led to outbreaks of skin rashes and nausea in West London, Exeter and Seaford, East Sussex. In all three cases, the authorities involved have done their best to cover up the incidents.

Doctors are slow to recognise the symptoms of agro-chemical poisoning, and for some of the chemicals there is no known antidote. An added danger is that many farmers — they readily admit to this — adopt a casual approach to 'interference', including instructions on how to use chemicals. They are reluctant to give information about where they store or dispose of chemicals, or how and when they are spraying from the air. They mix chemicals, in the interest of convenience and speed, although even the manufacturers do not know the risks involved. One agricultural worker in Kent (who prefers to remain anonymous) was covered with blisters after mixing an 'adjuvant' with two common pesticides, amino triazole and 24-D. The worker, according to Tony Gould, local organiser of the farmworkers' union, was provided with nothing more than a mask for use in a grain store. "It was a classic case", he says, "of a new substance being used, where the worker hasn't been properly instructed either as to how it should be used, or to what protective clothing he should wear". Nearly half the farmworkers questioned in a survey conducted by the Low Pay Unit in 1977 said they were not provided with protective clothing or equipment, a third had suffered at some time from skin irritation and allergies, and a third from chest and lung trouble. Farmers ignore dangers to themselves as well as to their workers, though as the Health and Safety Executive calmly puts it: "farms are not the places they used to be 20 or 30 years ago".

Chemical manufacturers proudly point to official statistics which show that there have been no fatal accidents attributable

Chlorinated hydrocarbons: DDT, aldrin, 24-D, 245-T	Effects, other than drastic, immediate effects, or long-term effects: Similar effects to solvents like chloroform — dizziness and headaches. Affects the nervous system. First symptom is restlessness and increased excitability, which may be followed by twitching — this can result in long-term liver damage.
Organic phosphates: parathion, diazinon, malathion, carbamates	Affects the nervous system — mild poisoning, mundane headache, fatigue and mild indigestion. May be delayed effects. Much research needs to be done to clarify behavioural symptoms of exposure.
Organic mercurials:	There has been a series of accidents due to seed dressings getting to the wrong place. Long-term effects not really known, although appears to effect brain functions — as in Japan, at the port of Minimata, where many people became permanently crippled.
Bipyridyl herbicides: paraquat and diquat	Acute dangers well known. Lower doses result in damaged skin and mucous membranes several days later. Attacks lungs, skin and fingernails.
Captan fungicide:	Causes gene damage — it is mutagenic.

Source: *The Landworker.*

to pesticides, and no more than 15 non-fatal accidents attributable to their products in any year since 1975. But cause and effect is very difficult to prove; we do not know how many incidents have not been officially reported or diagnosed, but we do know the difficulties in enforcing regulations in rural areas. Poisoned baits, for instance, are frequently placed so that their effect is indiscriminate, harmful not only to wildlife but also to humans, and to children in particular. The difficulty in enforcing the law is one reason why Governments have been so reluctant to introduce a system of licences covering the supply and use of agricultural chemicals. The Government's Pesticides Safety Precautions Scheme is voluntary — with the Ministry of Agriculture arguing that if it was made mandatory, more manufacturers would be encouraged to evade the controls they were told to abide by! Instructions, when they are spelt out in detail, are often very complicated — something which the manufacturers readily acknowledge.

The effect of chemicals in the food chain is undramatic in the sense that it may take several years before any concrete symptom appears. It was discovered — eventually — that DDT found its way into mothers' milk in the US. In 1978, 4,000 chickens died in East Anglia: it was discovered that sawdust in their litter had come from timber treated with dieldrin, a chlorine-based wood preservative. Yet owls had already died in London Zoo as a result of eating mice which had consumed identically-infected litter.

Chemical poisoning has contributed to the decline in numbers of more than 220 species of birds in Western Europe, with more than 50 threatened with extinction. Recent studies demonstrate that despite advice from the Government that DDT and dieldrin should not be used, more — not less — is being sprayed on crops, notably vegetables. Three monitored areas, Berkshire, Northumberland and Galloway, showed a significant increase in levels of DDE (a derivative of DDT) in sparrowhawk eggs. Earlier surveys demonstrated that, in spite of official appeals to stop the use in agriculture of organo-chlorine compounds, there was no fall in residues during the 1970s. The studies were dismissed by Whitehall, the agrochemical industry and the National Farmers' Union.

In theory, from January 1, 1981, under an EEC directive, DDT must be phased out: in practice, this is a voluntary agreement. Meanwhile, the Wessex Water Authority has expressed concern about the amount of dieldrin, which is still being used by farmers as a sheepdip, entering rivers.

Though banned for most uses in the US, persistent organo-chlorine pesticides are omitted even from the voluntary restrictions proposed by Whitehall. They are used on winter wheat, sugarbeet, barley, strawberries and, as we have seen, on vegetables.[3]

To sum up so far: the Government relies on voluntary codes of conduct and voluntary vetting procedures rather than statutory obligations; there is no independent long-term investigation into carcinogenic or teratogenic effects of agro-chemicals; little is known about the combined effects of a cocktail of different pesticides sprayed on the same product, or of the lasting impurities in pesticides; detailed studies on chemicals used on the land or in processing food are not published.

The Government and industry use familiar arguments: legislation would be time-consuming and costly; greater openness would provoke an emotional response among the general public, which in turn would either misunderstand the significance of the information, or be at the mercy of those who, for political motives, want deliberately to mislead.

The dramatic increase in aerial spraying — over one million acreas are sprayed from the air each year — illustrates some of the problems. Though farmers by law have to warn people and other farmers living nearby before they spray, this obligation is more honoured in the breach than in the observance. Neither the police, nor bee-keepers (bees are particularly sensitive to toxic chemicals) are generally informed. Even if they are notified in advance people are not told now to protect their gardens and animals. If they want to complain or seek damages, it is virtually impossible to establish a case without a clear admission on the part of the culprit. In this way, vegetable and fruit growers in the Vale of Evesham lose thousands of pounds worth of crops each year after aerial spraying of grain in neighbouring fields.

At the end of June 1980, the three children of John Hare

Chemicals Used in Farming

The Ministry of Agriculture's list of approved chemicals is now more than 300 pages long – a few years ago, it was a quarter of the size. Of the eight hundred compounds listed, here are a handful which have been linked to serious diseases or allergies:

2,4-D, a herbicide
Instructions: avoid spray drift, keep farm stock away until foliage of weeds has died, application to water courses may lead to illegal pollution.

This is sometimes mixed with 2,4,5-T and the two chemicals together made up Agent Orange used by the US in the Vietnam War. Agent Orange has been linked to birth deformities and other serious diseases affecting veterans of that war.

2,4,5-T, a herbicide
Instructions: same as 2,4-D. Used especially against brushwood and nettles.

Maleic hydrazide, a growth regulator
Instructions: similar (can be used with 2,4-D) with warnings about application to water courses.

This is used to suppress grass growth and is increasingly popular among local authoriities anxious to save money on less labour-intensive methods of cutting grass.

Paraquat, a herbicide
Instructions: harmful to animals.

This has led to a number of cases of poisoning. Developed by ICI , it has also been used by the Mexican Government against marijuana crops.

The Ministry of Agriculture agreed to go along with an EEC directive which, from 1981, imposed severe restrictions on the use of organochloride pesticides — persistent compounds which are particularly harmful to wildlife. They include aldrin, DDT, endrin and dieldrin. But the Pesticides Safety Precautions Scheme is voluntary. In 1969, the Government agree that the use of these chemicals "should cease as soon as practicable". They are still being used and neither Whitehall nor the companies release figures on the sales of pesticides.

The Ministry's list in 1981 included DDT with a long list of barley varieties on which it should not be used, and another which were "safe to treat". It also included chlordane, another persistent organochlorine, as a wormkiller in turf, though the EEC directive said it should be banned completely.

Aerial Spraying. *Aerial spraying threatens wildlife, neighbouring crops and people. Over 9 million tonnes of chemicals are sprayed on cereals alone in Britain every year. Some crops are sprayed as many as ten times. (Photo: Fox Photos)*

who lives in Benenden, Kent, suffered severe sore throats. Their doctor said that the cause was spraying of organophosphorus chemicals. The mixture was identified as Metasystox, officially recognised by the Ministry of Agriculture as harmful to humans as well as to animals. The local Health and Safety Executive in Ashford told Hare not to eat his garden vegetables for three weeks and to keep the family horse in the stable. The inspector reported the farmer concerned as saying that people living in the country must expect to be affected by spraying. This is just one case; the vast majority go unnoticed, unreported.

Aided and abetted by the Ministry of Agriculture, large agrochemical companies, the banks and high EEC prices, farmers have rushed to produce as much as possible. If they run fast enough, they have been told, they will eventually overtake the rising costs of their fertilisers, pesticides, fungicides, machinery and the interest on their overdrafts. Chemicals, the companies say, are needed to help save the millions of hungry people in the world and one-third of the world's harvest is destroyed by pests. Civilisation needs pesticides, says the National Farmers' Union. "The balance of nature is a culling process against civilised man", according to Dr Hessayon, a leading member of the British Agrochemicals Association and chairman of Pan Britannica Industries Ltd. Nature has time for the fungi and the flies, but has little time for man.

It is not suggested here that the use of all chemicals in any quantities is always dangerous and should be stopped immediately. But there must be serious concern about the motives behind the vested interests as well as the evidence that exists about the damage chemicals inflict. The harmful effects, the priority given to short-term profit over longer-term security and health are actively promoted by the Ministry of Agriculture, which as a result is working directly against the real interests of agriculture.

Pesticides are one of the fastest growing sectors of the chemical industry. Sales in Britain over the past few years have risen by more than 30 percent annually, and Shell confidently predicts that the world pesticides market will increase from $4,000 millions in 1975 to $8,600 millions in 1986. It costs between $8 millions and $12 millions to develop a new pesticide. There are already 800 different types, with 9.5 million tonnes sprayed on cereals alone in Britain in 1979.

Encouraged by the agrochemicals industry and the Ministry of Agriculture, farmers drench their fields: a crop of potatoes, for instance, can be sprayed as many as 10 times. Many farmers have adopted the attitude that "what's good for agriculture is good for the environment". Others, as we shall see, are becoming more sceptical, but as the Royal Commission on Environmental Pollution reported: "we have been told by some farmers that they feel themselves to be 'on a treadmill' with regard to pesticide usage, compelled to depend on chemicals to an extent which they, as countrymen, intuitively find disturbing".

As colour advertisements in the agricultural press constantly repeat, it is a sign of poor management to have a single wild oat popping up over a field of barley. Farmers prefer to over-compensate, "to play safe" (sic). According to the British Crop Protection Council, "more than one million times as much pesticide may be applied as would be needed to kill the pests". Dr George Cooke, chief research scientist at the Agricultural Research Council argues that the quantity of pesticides used by farmers could be reduced 1,000 times and still be more effective with more appropriate, less wasteful, but more sophisticated spraying techniques.

Many chemicals (and probably half the 3,000 additives used in processed food) are applied merely for cosmetic reasons. It is a practice deliberately encouraged by the agricultural pricing policy. For example, the profit lost by selling grade 2 apples with a few minor blemishes instead of grade 1 fruit would be more than the cost of applying 78 applications of pesticide to ensure that the grower obtains the higher grade 1 price. The Royal Commission on Environmental Pollution suggests that the Food and Drugs Act should be amended to reduce the pressure on food producers to aim at absolutely pest-free fruit and vegetables. Such an amendment could make it a defence to show that any extraneous material was not hazardous to health.

The chemical companies quickly seized on the suggestion, claiming that consumers expected unblemished produce. The Ministry of Agriculture, appealing to British growers to market their produce more aggressively, quotes the Dutch and French whose fruit and vegetables are of a constant, predictable, size and colour only because they have been dosed in chemicals. Would people really object to buying cheaper, tastier, more natural, produce even though it may have a slight blemish, asymetric shape, more than one shade of colour?

For farmers, the most serious problem in the fight against pests is the build-up of resistance. Insects are now the only species, with the exception of man, which is actually increasing in numbers. There are about 270 kinds of insects that are immune to most pesticides. A resistant strain of aphid is developing with extra genes and more enzymes which break down insecticides. Ordinary flies are developing a blanket resistance to insecticides, causing problems to intensive live-stock units but also a potential public health risk. The more resistant the insect, the stronger and more toxic is the chemical used to counter the problem. As insects became resistant to organochlorine insecticides, they then develop resistance to organophosphorus ones invented to replace them. And the less biodegradable — the more persistent — the insecticide, the easier it is for the insects to develop resistance to them.

Farming is becoming heavily dependent on the chemical industry. But the industry is quick to reply. Spokesmen for the British Agrochemicals Association argue that malaria has returned to so many areas of the world because DDT has been banned by environmentalists. More to the point, as the World Health Organisation points out: "the resurgence in global incidence of malaria over the past fourteen years can be attributed very largely to the resistance of anopheline mosquito vectors to previously effective insecticides, though lack of funds has contributed to the failure of some eradication programmes. Agriculture is often both a major contributory cause of this resistance, through the widespread use of insecticides against crop pests, and the chief sufferer from the disabling and weakening effects of the disease". Mosquitos resistant to both DDT and dieldrin are now breeding in Africa, Central America and parts of the Indian sub-continent and the Middle East.

There is a real danger that pests will eventually cause more damage than before chemical controls were applied in the first place, and that with natural predators dying out and with new mechanical cropping methods relying on chemicals, habitats will become ideal breeding grounds for pests. But the Ministry of Agriculture merely expresses the view that resistance is unlikely to be a major constraint on agricultural production by the end of the century. It also argues that the introduction of stricter controls would lead farmers to seek compensation, either in higher food prices or in extra subsidies from the taxpayer. This is an archetypal Whitehall response, though no less extraordinary for that. For there is ample evidence that a reduction in the use of chemicals is not only cheaper and better for the soil but will also give the farmer greater profits.

What is certain to become an increasingly dramatic and crucial problem was illustrated by Sir Emrys Hughes, then principal of the Royal Agricultural College, Cirencester, and a pillar of the agricultural establishment. He estimated in 1978 that the cost of growing cereals over the previous five years had increased four-fold. It was going to cost £118 an acre the following harvest, with the result that virtually the whole of the first two tonnes per acre would be needed merely to cover expenses. "I think that the farmers in the UK, as a whole, are on this treadmill," he said, "and there is an air of desperation in British cereal farming which I sense at the present time. Ten years ago it was the 3-tonne club in Lincolnshire, now it is the 10-tonne club". "But you see," he added, "farmers did get 5 tonnes per acre last year. I went to see some of them and one cereal grower had made 15 passes across each of his fields; he had applied something 15 times; four split dressings of nitrogen, all sorts of herbicide cocktails of one sort or another and so on. But he has got 5 tonnes . . . We don't put 15 dressings of herbicides and other chemicals on a crop of cereals without something happening other than increased yields. We are changing the whole of the ecosystem; we are bound to be. Two years ago one of the great frights we had at the College farms was aphids; aphids on cereals, that is . . . I had never met them before in my 50 years of farming . . . suddenly they can destroy a cereal crop".

Contrast that warning by Hughes, echoed by a few other agricultural scientists — but by many more concerned individuals readily dismissed as 'eco-freaks' — with the report by the National Economic Development Council on Agriculture into the 1980s. It stated: "the production of agrochemicals is mainly in the hands of international companies dealing in high-value products which are the result of intensive research and development . . . it is important to avoid any unnecessary constraints on agricultural output through limitations on the use of agrichemicals".

We need not discuss here the moral issues which have helped to make factory farming such an emotional minefield, but I will consider the causes and effects of this essentially industrial system; it has little to do with agriculture. It was the drive to produce as much food as possible as cheaply, as quickly and as profitably as possible that led to factory farming. But as part of this drive in the name of efficiency, at a time when some 3,000 animals are slaughtered every minute of every working day, the accent is also on uniformity, a process which treats animals simply as production units, even to the extent, for instance, of depriving sows of that most natural task of suckling their own piglets.

William Archer, a veterinary surgeon, wrote in 1976: "farming economics, commercial greed and establishment indifference have combined to produce conditions in modern farming which are, frankly, disgusting. So-called welfare control in the form of visits by Ministry of Agriculture veterinary staff are known, by all concerned, to be a joke.

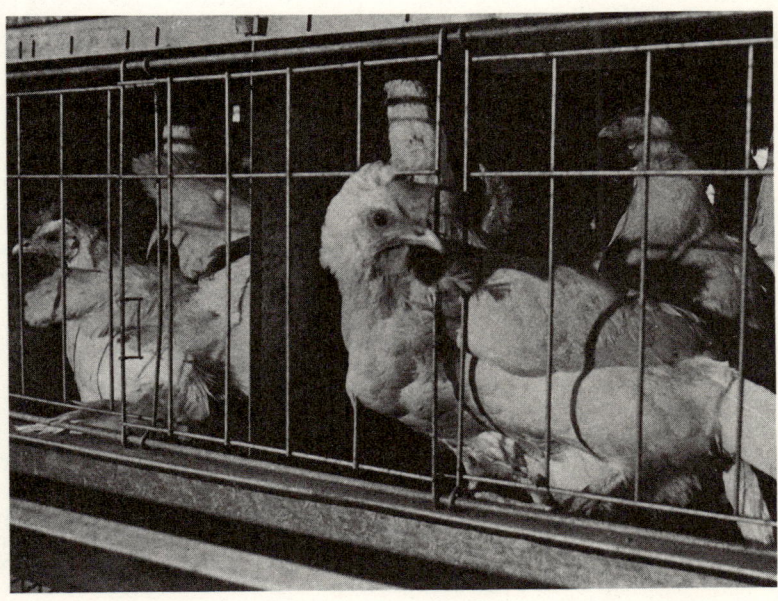

Battery Hens. *They cannot fly, scratch, spread their wings, have a dust bath. Hens are bred specially to survive in battery conditions and to consume enormous amounts of high energy rations. With three square feet per bird – a fraction of the space they have in battery cages – they would use a very small amount of the acres now needed to grow concentrated feed for them. (Photo: The Guardian)*

Statistics presented to Parliament are, to be generous, mis-leading. Codes of practice produced by politicians' servants are intended solely to lull us into believing that factory farming is pretty and the animals therein protected. Anyone expressing concern for modern factory farming methods is dismissed as a crank". The law says that animals must not be made to suffer "unnecessary pain and distress". It is full of loopholes; who is to decide and on what criteria? The 1975 Brambell Committee recommendations — the official yardstick — are widely ignored, or as official spokesmen put it: "the industry has been slow to respond to them".

That Committee, appointed by the Government, said that the debeaking of hens (still widely practiced) is "cruel and objectionable". It said that cages for laying poultry should contain no more than three hens and measure at least 20" wide, 17" deep and 18" high. Four birds per cage — smaller than the dimensions recommended — is common and five per cage is not infrequent. Chickens are kept in windowless barns, allowed just 4" width each though their average wing span is 32". They cannot perch, fly, scratch, spread their wings, have a dust bath. Hens are being bred specifically to survive in battery conditions and to consume enormous quantities of high energy rations.

Breeding sows are chained to straps in steel farrowing crates, unable to turn around, sometimes for the whole of their four-month pregnancy. Calves are often kept in small crates in which they are unable to turn round and they are deprived of normal access to water, solid food and straw.

As a result, pigs suffer from stress, hens become cannibal-istic, partly through boredom. In some poultry stations a death rate of one in ten is considered normal and acceptable. Defenders of the system argue that less intensive systems provoke fights for the pecking order, stress, cannibalism, threats from predators. But at least there is little dispute that an unnatural environment sharply increases the incidence of disease. There has been a significant increase in tumours, respiratory diseases, mastitis in cattle, pneumonia in cattle and hepatitis, in intensive units. Piglets are sometimes removed surgically from the womb of sows and taken to sterile places to avoid the risk of virus pneumonia, itself a disease provoked by intensive farming. That factory farming has increased the likelihood of disease is no longer questioned. "Salmonella infections have been known for a long time", said the British Association for the Advancement of Science in 1978, "but there is little doubt that the increasing incidence of the infections of livestock (particularly in pigs and poultry) is linked with the intensive systems of livestock management which have been developed over the last two or three decades".

And for those who work in intensive livestock units, it is a very unhealthy and dangerous existence. Tenosynovitis, the inflammation of the wrist tendons caused by rapid repetitive movement and common among coalminers in the past, is a well-known hazard in poultry factories. Newcastle disease, an eye infection, is carried by the dust from chickens' feathers. Working with the noise and smell of factory farms is an experience closer to a chemical factory than anything conjured up by advertisements on behalf of the food industry.

The Commons Agriculture Committee, grouping back-benchers from both main political parties, said in July 1981: "We do not accept the contention, frequently stated or implied, that the public demand for cheap food decrees that the cheapest possible methods of production must be adopted . . . Society has the duty to see that undue suffering is not caused to animals, and we cannot accept that that duty should be set aside in order that food may be produced more cheaply. Where unacceptable suffering can be improved only at extra cost, that cost should be borne or the product foregone".

After an exhaustive inquiry into animal welfare it added: "We have gained the impression that within (Whitehall) Departments the whole weight and thrust of policy, until recently at least, has been directed towards ever-greater productivity and profit, and that the welfare of the animals concerned has played at best a minor part in Ministerial and official thinking". It said that there was a feeling that animal welfare was still regarded as "a tiresome complication engendered by vocal sentimentalists who need to be placated at minimum cost to producers' profits. To the extent that this is true, it is high time for a change of attitude".

It is no accident that British standards of food hygiene —

and public health in general — are lower than in most other Western countries, contrary to what popular prejudice would have us believe. There is no strict system of reporting cases of salmonella poisoning of either animals or humans, while regulations covering heat-treatment of animal feed, belatedly introduced in Britain, are respected only in a haphazard way. In a study sponsored by the Public Health Laboratory Service in 1972 twenty-three percent of meat and bonemeal samples in Britain contained salmonellae, whereas only 0.3 percent of Danish heat-treated produce had been contaminated. Salmonellae, and more and more sophisticated types, are on the increase. "Quite frankly", says Dr George Gibson, public authority chemist in Leeds, "we find salmonella wherever we look". The number of deaths attributed to salmonella infections among males of working age rose progressively from 3 to 4 percent a year in 1968-69 to 15 percent a year in 1975-76. In an indictment of food processors, the Association of Meat Inspectors in 1979 delivered a blistering attack on the state of British abattoirs, only a third of which, it noted, were meeting even basic hygiene standards. The meat industry itself acknowledges that the transport of animals frequently causes unnecessary suffering and that stunning methods are ineffective.

The use of hormone drugs, antibiotics, and anabolic steroids to promote growth, and weight by increasing the amount of water in the animal in pigs, poultry and cattle (about a fifth of Britain's cattle herd is treated with growth promoters) also encourages disease and resistance to conventional treatment. Poultry in particular suffer from heart attacks simply because they are growing too fast. After 100 chickens had been stolen in Peterborough in August 1978, the police warned the public that if eaten, the carcasses, which had been treated with a hormone drug, could be dangerous. Farmers spend over £20 millions a year on antibiotics alone.

Producers of veal got a shock in 1980 when consumers in France and Italy cut their purchases by a half after fears were expressed about the side-effects of hormone residues. Although they are officially banned in both countries, the law is not enforced. The EEC Commission came up with a proposal for strict controls over the use of natural and synthetic hormones and anabolic steroids, an initiative which caused great dismay among senior officials in Britain's Ministry of Agriculture on the grounds that Governments across the Channel had allowed themselves to be persuaded by "public emotion" rather than "independent scientific investigation". The attitude in Whitehall reflected the concern over the controversy over 245-T.

For the record, growth promoters permitted in British turkeys include: payzene, zinc bacitracin, virginiamycin, flavomycin, 3-Nitro, arsanilic acid, grofas, and other additives, including blackhead preventatives, allow for 105 possible combinations.

The farming establishment in Britain was also shaken a year earlier, in 1979, when the High Court in Frankfurt ruled that it was cruel to keep hens in battery cages since it deprived the birds of the ability to pursue instinctive behavioural patterns including scratching, stretching, flapping their wings and preening. The ruling brought pressure on the West German government to combat the worst examples of factory farming throughout the Common Market.

Factory farming is more akin to an industry than agriculture, but it escapes planning controls. Most of the 60 million tonnes or so of slurry which factory farms release every year remains untreated and there are no adequate guidelines on how and where it should be disposed. Factory farmers do not pay the cost of pollution control, like other industries, and local authorities have no power to enforce controls. Humberside county council has set an example by drawing up a Subject Plan on Intensive Livestock Units, the first local plan to deal specifically with agricultural development. It was prompted by concern over pollution in an area where there are more pigs than in any other county in England and Wales. But the council still cannot get round the problem that intensive livestock units can be classified as 'agriculture' and, as such, be exempted from traditional planning controls.

Defenders of factory farming say that alternative systems would need more land, more labour and would result in more expensive food. Yet there are 4 million unemployed. Eggs, the National Farmers' Union estimates, would be 80 percent dearer if free-range systems only were allowed. Yet the criterion for this argument is the conventional assumption about what is 'cost-effective'. The cost of maintaining artificial lighting in intensive pig and poultry units is extremely high. Yet natural light is free. With three square feet per bird, the entire country's flock of hens could be housed on 12,000 acres at most. a fraction of the 2½ million acres now needed to grow concentrated feed for them. Free range systems — almost all hens were free range fifty years ago — would produce good manure for the land.

There are chinks of light at the end of this particular tunnel. Volac, Britain's largest veal producer, has developed a system whereby calves live in groups of 30 in straw-filled pens. Less capital equipment is needed and less artificial lighting. Loose-housed calves won all the prizes at the 1979 Smithfield Show in London — some slight comfort, perhaps, for the vegetarian groups who were demonstrating outside. (It is worth pointing out that while Britain's export trade in calves has been booming over the past few years, we import 90,000 carcasses of veal from the Continent annually.) Does this make sense?

Even the Ministry of Agriculture is now investigating alternative poultry farming systems with the 'aviary method' which allows hens to roam freely. Research at Cambridge by Dr David Sainsbury of the Department of Clinical Veterinary Medicine into a straw-yard system has shown that hens can lay as many eggs as in the battery-cage system and at the same time eat less concentrated, and so cheaper, food. A small amount of electricity would be needed, maybe, to stimulate laying on dark winter mornings. Otherwise, Sainsbury concedes only that the system would require more labour for the collection of eggs, and the feeding and watering of the birds.

Pig farmers are beginning to appreciate that free-range systems, with small and cheap, corrugated iron shelters dotted about a field, are as profitable as intensive-rearing methods.

In extensive mixed farming systems with livestock fed with a normal diet, manure can be used on the land. Intensive farming poses the threat of contaminated sewage, or raw sewage sludge. The Environmental Health Officers' Association warned in its 1978 annual report: "by spreading poultry manure on grazing land it seems inevitable that sooner or later the good work done during the Tubercolosis Eradication

Scheme will be lost with the re-introduction of tubercolosis from avian sources". It also suggested that the increase in cattle diseases appears to be linked to the increase in the amount of sewage sludge and slurry spread on grazing land.

An indication of how Whitehall treats the problem was given by Kenneth Marks, then under-secretary at the Department of the Environment, after 30 cows had died from fluorosis in a Yorkshire farm in 1977. "My Department's guidelines on sludge disposal do not refer to a limit or analysis for fluorine because it has not been regarded as a problem in the past", he said. There had been little or no research into the effects of the disposal of sewage sludge on agricultural land and no effective monitoring of toxic elements in sewage sludge.

The viscious circle in which we are trapped is illustrated by the vulnerability of high-yielding wheat to fungus disease. The seeds need ammonium nitrate for protection (about one million tonnes are spread on the land each year). But nitrate encourages the spread of fungus disease. Not only that; nitrates can cause disease in humans is it enters the water supply. The level of nitrates in London's drinking water has already exceeded the safety limits recommended by the World Health Organisation. But as one writer has put it: "according to the Thames Water Authority this is not a matter for concern because it can be dealt with in future by adding another chemical to the drinking water". Nitrates in water caused a particular serious problem during the long drought in 1976.

What are the alternatives? To begin with, without even looking at the causes, at least half Britain's pig population could be fed on treated, processed animal feed wastes, saving one and a half million tonnes of grain. (With the assistance of slugs, snails and worms, vegetable waste — residues worth £100 millions a year — could also be turned into animal feed.) Liquid slurry from cattle units could be turned into natural gas, with some left over to be used as dry compost. Biogas plants — using manure and crop wastes to produce methane and with carbon dioxide added — are used extensively in China for heating. The Welsh Water Authority has used sewage waste to turn sand dunes into rich grassland; fiveacres along the south Pembrokeshire coast have been fed on two million gallons of treated sludge. An attack on waste has inspired local authorities in Bedfordshire and Eastbourne, a conference centre town with an exceptional supply of waste paper, to turn matter into soil. But these are exceptional examples; for the most part, farmers and their advisers, local authorities and central Government still adopt a complacent attitude towards waste, clinging to the traditional, and somewhat paradoxical argument that recycling and stricter control over waste is too expensive.

Avoidable disease is a waste, so is the practice of taking nutrients out of processed food, like white bread and fattening artificially cattle and sheep. Poor transport facilities for livestock is a waste since it causes stress thereby spoiling a great deal of the carcase. It is a waste, too, to feed animals with expensive concentrates; a fall of 1 percent in the amount of feed given to livestock would save £18 millions annually. Highly nutritious but aesthetically unattractive food such as black pudding or tripe is wasted; barely half the weight of fish landed at the quayside is actually eaten. On average, about 25 percent of all food, from the time it is harvested to the moment it arrives on the dinner table, is wasted.

It is scarcely surprising that technology, which has led to so much waste and pollution (and has also brought immense benefits in the fields of health, safety and medicine — the author does not want to inhabit an artificial Stone Age) has also developed ways of using the waste constructively. Straw and manure can be used in feed, rich in protein. But in any study of agriculture, the land, and the environment, it is difficult not to come away with the impression that for every step forward, we are taking two steps backwards. We are encouraged to adopt an attitude of mind which makes us lazy, selfish and short-sighted. Ask yourself: is it not wasteful when buses with just a handful of people are stuck in traffic jams caused by hundreds of private cars with the driver, but no-one else in them?

What is the connection between saving some of the most beautiful areas of Britain and pulling the chain in the lavatory? In his campaign to save Wastwater from British Nuclear Fuels (which wants to take 11 million gallons a day from it) and Ennerdale water from the North West Water Authority which wants to take 12 million gallons extra a day, Christopher

Brasher has pointed out in *The Observer* that each one of us uses about 35 gallons of clean, drinkable water every day. Twenty-five gallons of this (the equivalent of 1,400 million gallons a day) is used in flushing the lavatory and washing and bathing. By filling a large plastic bag with gravel, tying the ends, putting it in the cistern, you will save about three gallons a day, over 8 percent of all the water you use, he has calculated. Of the 12,777 gallons of water consumed per person per year, 4,560 go down the lavatory, and just 365 gallons are used for drinking and cooking.

We waste hundreds of thousands of tonnes of raw materials — sand, limestone and soda ash — every year as well as energy, ratepayers' money through waste disposal by indulging in throw-away containers: bottles, cans and jars. Half the cost of a can of drink is the cost of the container: if they were refilled, consumers would save more than £100 millions a year, the Friends of the Earth estimate. Beverage containers account for 40 percent of the contents of the average British dustbin. A returnable container system would create jobs and save energy costs. A Bottle Act is a blindingly obvious need.

Organisations like water authorities or energy producers, whether the nuclear industry or the Electricity Council, have a vested interest in telling us we need to consume more and more. They want larger budgets, attract the attention of Whitehall, have a greater say and become more important. They are no different in this than the agrochemical industry. They want us to consume as much as possible. They will persuade the Whitehall bureaucracy, which, like any bureaucracy, is always concerned about being 'caught out', to 'play safe', they persuade it to accept demand forecasts which are far too generous.

Largely because of the economic climate in Britain, total energy demand has remained virtually constant at 350 million tonnes of coal equivalent a year since 1973, a fact studiously ignored by the energy industry. The Atomic Energy Authority and the Central Electricity Generating Board jealously guard technical information over which they have monopoly control. The CEGB, for example, justifies its pro-nuclear position by referring to what it calls "net effective system costs", a narrow techno-economic criterion which claims that even without its

Torness Protest. *Anti-nuclear protest at Torness, East Lothian in 1978. Scotland already has sufficient electricity supply to cope with 70 percent more than existing peak demand without the nuclear power station now being constructed here. (Photo: Camera Press)*

forecast growth in demand new nuclear power stations would be more cost-effective and cheaper to run than existing coal or oil-fired stations.

And so the Government intends to construct at least 10 new nuclear power stations over the coming decade. Scotland already has sufficient electricity supply to cope with 70 percent more than current peak demand even without the nuclear power station at Torness now being constructed. Why does the North of Scotland Hydro-Electric Board need to invade Loch Lomondside? "Their scheme", says Brasher, "involves using electricity in the small hours of the morning to pump water uphill for about 1,500 feet into a new reservoir built in the Shady Glen on the steep slopes of Ben Lomond. Then at tea-time every day, or just after the crowning of Miss World, they can let the water down through massive turbines back into Loch Lomond". Access roads will be dug — just as they have been for similar reasons near Llanberis in Snowdonia — through unspoilt countryside. The Hydro-Electric Board was, of course, secretive about the cost of the scheme and refused even to inform the chief district planning officer about the plan. It refused to disclose the contents of a 158-page environmental study it had drawn up.

We waste water because we do not think, or do not care, about the implications of our acts. And it is our approach to energy which is most obviously wasteful, potentially disastrous and most harmful to the land, our most precious and renewable resource.

The countryside and the land are threatened by the search and insatiable appetite for energy sources — coal, electricity, nuclear plants, oil, water and minerals. According to Sir Martin Ryle, the Astronomer Royal, the capital cost of a nuclear power station, if spent instead on energy saving, would save three times more energy than the station would produce in a lifetime. The ten nuclear power stations the Conservative Government wants to build will provide only about 10 percent of our energy by the year 2000 on current forecasts. Yet effective insulation could cut energy consumed in all buildings by 50 percent, or the total energy bill by 17 percent. Three-quarters of the heat produced by coal-fired electricity power stations is wasted. The gas equivalent of two days' supply of British North Sea oil goes up in flames every day.

Government home insulation grants were one of the first casualties of the public expenditure cuts of 1979-80. According to a confidential memorandum drawn up by Whitehall officials and sent to David Howell, the Energy Secretary in February 1980: "conservation is now well down the list of priorities of other Departments if it is on their lists at all". Yet Britain is in danger of developing a more energy-dependent economy than any of its main competitors.

Increasing reliance on energy-intensive capital is one of the hallmarks of agricultural development in Britain and, indeed, in the rest of the world. Moreover, this dependence has one main victim: labour. Yet the Ministry of Agriculture persistently argues that "British farming is uniquely efficient". This is true only if you take the yardstick of output per man. John Bowman, former director of the Centre for Agricultural Strategy, has calculated that while in terms of labour productivity, agriculture comes top of a list of 17 main industries, in terms of capital employed it is seventh, and in terms of energy use it comes eleventh.

Modern farming is not an efficient converter of fossil fuel energy into food. "Our present farming methods", Sir Kenneth Blaxter, one of the country's leading agricultural scientists, told the 1978 Oxford Farming Conference, "have developed because oil has been cheap". For every calorie of food produced from sunshine, he added, Britain uses ten from residues of plants fossilised millions of years ago; for every unit of energy we eat, we expend nearly ten units of fuel when processing, storage and distribution is taken into account.

There is probably not a single crop in the West which produces net energy. The amount of energy needed to produce one tonne of nitrogen fertiliser is equivalent to the amount of energy you get from burning five tonnes of coal. To put it another way, each tonne of nitrogen needs 1.8 tonnes of oil to produce. On this basis one person's ration of wheat a year uses in fertiliser 2 gallons of oil. About twenty percent of the country's total fuel consumption is used in getting food from the farm to our dinner tables (and we have already noted that the distribution of food is the biggest single user of commercial traffic on the roads).

As Blaxter pointed out, the change from horses to tractors this century has increased the power used by farmers sixteen times and energy used to provide fertilisers and weed control chemicals has risen by 20 times. When agriculture depended on horses and labour, Britain was 41 percent self-sufficient for a population of 38 millions. Now it is about 46 percent self-sufficient, expressed in terms of food energy, for a population of 56 millions.

Fossil fuel consumption, according to the Ministry of Agriculture's own figures, increased between 1942 and 1972 3.4 times for an increase in wheat yield of 2.1 times.

Over the past twenty years yields have increased substantially, especially in wheat and dairy farming. But while the

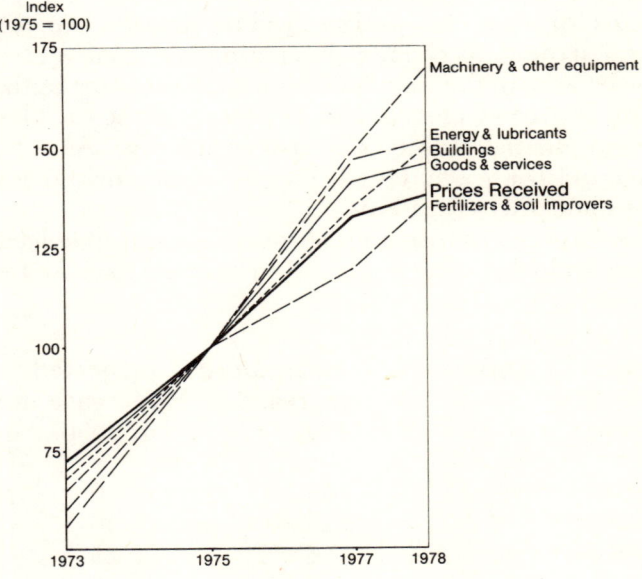

23. **Agricultural Costs and Incomes:** *Many farmers are on a treadmill, desperately trying to cover rising costs by producing more food as quickly and intensively as possible, which means increased dependence on chemicals. They "save" most on labour costs, which now account for only about 16% of total costs. (Source: M.A.F.F.)*

rise in dairy cow yields is mainly due to the increased consumption of concentrated feed, more than half the increase in wheat yields had been the result of what the National Institute of Agricultural Botany describes as "improved varieties". Increased yields cannot simply be put down to better husbandry.

Britain has replaced low cost, low risk, low yield agriculture with high cost, high risk, high output, energy-consuming agriculture. And there is growing evidence that farmers have reached the point where the law of diminishing returns takes over. Farmers' prices in Britain are among the highest in the world, yet farmers cannot keep up with their rising input costs; of fertilisers, chemicals, machinery and fuel.

Blaxter argues that for wheat, barley, sugarbeet, eggs and milk, the annual increase in yields has been declining and that without a new technological breakthrough the present impetus for these commodities will be virtually expended by the end of the century. He estimates that the mean increases in yield per acre in the year 2000 would, at the most optimistic, remain between 5 percent and 30 percent of the 1947 level. In addition, the net output of British agriculture seems to have peaked in 1974-75, and totalled only 80 percent of that value, at constant prices, in 1976-77.

"Attempts to reduce animal mortality, improve feed conversion and raise crop yields still further are likely to be both

product	unit	mean annual increase in 1947	projected annual increase in yield in 2000.
wheat	kg/ha	133.30	15.10
barley	"	123.30	6.30
sugarbeet	tonnes/ha	1.33	0.25
eggs	no/bird	6.60	1.00
milk	litres/cow	67.70	18.60

source: Blaxter, "The use of Resources", *Hammond Lecture to the British Society of Animal Production, Harrogate, March 1976, and reprinted from "Land for Agriculture", CAS Report 1.*

more costly and less rapid than in the past", the CAS added in its report. "Potential increases in output per unit area will probably be more dependent on *human* effort and attention than previously".

Ironically, the highest yield recorded in 1977 — 4½ tonnes an acre of a new dwarf wheat, Maris Hobbit — was achieved by a farmer in Essex who used the traditional but increasingly ignored rotation method.

In spite of an average price increase of 34 percent between 1975 and 1978, earnings for British farmers during the period fell behind their costs: oil and energy prices increased by 170 percent between 1973 and 1978, fertiliser costs rose by 131 percent and maintenance and repair charges by more than 100 percent, according to Ministry of Agriculture figures. Even ICI acknowledged by 1979 that farmers must reduce their costs, adding blandly that they must also increase their production. "There is no other way he can survive", said ICI's chief agriculturist, David Lewis.

The British cow eats on average more expensive concentrated feed than her cousins in other EEC countries. Feed costs are about the same as on the Continent, but there family labour compensates. "In these circumstances," John Cherrington, successful Hampshire farmer and the respected, maverick, agricultural correspondent of the *Financial Times*, has put out, "economies of scale are largely worthless". If British cattle relied on grass alone, milk yields may fall by about a third, roughly to the average level in Ireland. Yet Irish farmers, resolutely stave off attempts at seduction by representatives of the large agrochemical and feed companies. In spite, or because of this, net profits per acre are higher than in Britain.

The National Agricultural Society of England recommends that farmers should keep their cows indoors throughout the year — never allow them to eat grass out in the open — since this would lead to more predictable yields. Yet the Milk Marketing Board, not noted for its radicalism, warns that intensive dairy farmers, at the rate their costs are increasing, would be losing money by 1984, and that the most intensive their methods, the more they would lose. Their prices will not increase because the market for their produce is simply not there. The MMB conducted two surveys which showed that while Irish and Breton farmers produced less milk from their cows, they enjoyed much higher profits because they fed their animals with less expensive concentrates, and relied much more on grass and clover. It concluded its report on Brittany by saying: "perhaps the one question above all others that the studies pose is whether modern milk production really needs to be a complex business, requiring ever-increasing levels of automation".

While millions of pounds each year are channelled into research into cereals, a basic resource such as grass is neglected. Yet Britain has an abundance of grass, the weather is perfectly suited to its growth. Recent research has shown that the heavy use of nitrogen fertilisers on grazing land makes the grass much more vulnerable to pressure from the weight of stock and machinery and kills off the rich variety of plant life in traditional meadows.

Types of grass that will respond to high nitrogen inputs have been carefully and expensively developed, yet clover could be just as productive. Nitrogen actually squeezes out clover, yet clover is a good and natural fixer of nitrogen. While farmers in Australia and New Zealand are being encouraged to seed mixed swards of clover and grass, farmers in Britain are applying over a million tonnes of fertiliser on grass at a cost of over £150 millions a year.

It is more than likely, if the agricultural establishment does switch its concentration to grass, it will encourage uniform types and urge farmers to plough up all the remaining old meadows and pastures that provide the tough, natural plants which are not only more than adequate in themselves, but also provide the characteristics needed for the development of new varieties.

Even the Ministry of Agriculture, when asked by the Royal Commission on Environmental Pollution to prepare a report on the possible effects of substantial increases in energy costs by the end of the century, acknowledged that less energy-intensive systems "might result in a shift, reversing trends, towards mixed farming patterns". A return to mixed farming, it added, would also have the advantage that a higher proportion of feed could be produced close to livestock.

Mixed and more extensive farming methods would involve more labour; not a bad thing, one would have thought, with unemployment now over the 4 millions mark. Yet the National Farmers' Union says as a criticism of those who oppose factory farming that alternatives would require 14,000 additional skilled stockmen. Even the International Agricultural Federation admits that "biological farming has the potential to provide abundant and healthy food at reasonable prices. It adds: "Success requires skills far above average and training which is not available".

But above all, it needs commitment and a change of attitude. The general approach of British farmers towards training was perhaps best illustrated in a survey sponsored by the Agricultural Training Board in the late 1970s. Less than 15 percent of school-leavers entered farming through the official apprenticeship scheme, and only one third of the apprentices were completing their three-year training period. Low pay was a major cause; but the survey, conducted in 1979, also showed that many farmers were simply not interested in cooperating. One apprentice explained: "Well, I got the sack. My employer said I was the most careless person he had had on the farm. I had two accidents involving machines. A band sprayer for sugarbeet and another sprayer where the hydraulic coupling got caught up on the tractor and broke. I tried to explain to the farmer that I was tired from working long hours. I went to the farm as a machinery apprentice but I just operated them. They didn't explain how to use them or how they worked".

Another said: "Well, I found it boring. It was too complicated and the things they told us were of no use. They were all scientific. I could never put them to use. The teachers didn't pay enough attention to what we were doing on the farm . . ."

We have seen how specialist farms are wasteful and face ever-increasing costs. It is also now widely accepted that small farms can be as productive per acre, and certainly as efficient in terms of energy, than large ones. Small, family, farmers can harvest ten times as many calories of food energy for every calorie expended on labour while it is estimated that in the US up to 20 calories of energy are consumed to produce and distribute a single calorie of food — expressed in terms of energy — to the consumer. Even US Department of Agriculture studies have suggested that small, labour-intensive, farmers are more efficient than larger ones.

So the agrochemical industry has to rely on the argument that not everybody wants to be a small, family, farmer and that fewer chemicals means more ploughing by tractors. Yet the damage inflicted on the soil by modern, conventional, agricultural methods is frightening. The reduction in organic matter upsets a complex biological system that has unfolded literally over millions of years. It upsets the balance of trace elements and small, delicate, flora and fauna on which the fertility of the soil depends. Agriculture, after all, is supposed to be a biological science.

The most dramatic, and most frequently quoted, example of maltreatment of the soil is the series of dustbowls in the American mid-West during the 1930s. The mid-West, where farmers have been under less pressure than British to squeeze maximum yields out of their crops, is also becoming less productive. Underground water supplies — aquafers — are being used up so that farmers have to dig deeper wells with the help of ever more energy-intensive pumps.

More than 50 million acres of land in the US are in poor condition because of over-grazing, according to a recent Federal survey. After a study of nearly seventy farms in twenty-three states, the US Department of Agriculture in 1981 described the heavy dependence on energy and expensive chemical fertilisers, a steady decline in soil productivity because of soil erosion and loss of organic matter and human and environmental hazards from chemicals as "the adverse effect of our US agricultural production system".[4] Organic farmers, it added, "have not regressed to agriculture as it was practised in the 1930s. They do use modern equipment and new crop varieties".

Organic farms, the report found, were almost 2½ times more productive per unit of energy consumed than conventional farms. Short-term profits were lower but if long-term costs, including soil erosion and water pollution, were taken into account, organic farmers were more competitive.

When a serious drought hit the corn belt in the mid-1970s, for instance, organic farms recorded equal or higher yields. A study by Washington University found that organic farmers used only 40 percent of the energy needed by conventional farms.

But in East Anglia, too, the destruction of hedges and a sharp fall in the organic content of the soil has resulted in the topsoil being blown away in many areas. In clay soil regions around Warwick and Northampton, after fifteen years of intensive, mechanised, one-crop agriculture, the soil has deteriorated to the point where two wet seasons make it difficult to continue with arable farming.

The Agricultural Research Council has discovered that direct seed drilling (i.e. without ploughing) on heavy ground weakens the soil, causing deep cracks. ICI developed paraquat to avoid the need to plough. Government scientists then spent years and thousands of pounds of public money trying to solve the problems of excessive soil compaction, disease control and poor crop performance resulting from the use of the chemical. Then ICI announced a new chemical cultivator to clear the new obstacles. It is a classic example of looking for a cure for new problems — an approach which is profitable for the chemical companies and the researchers, many of whom are financed by big business — rather than seek out the cause.

As soil is harmed so, too, are animals. Cattle breeding societies are concerned about the use of steroids and hormone growth promoters in bulls, and even the Meat and Livestock Commission admits that hormone treatment is probably a waste of time. Research has revealed a drastic fall in fertility of bulls fed on intensively-fertilised fodder.

Organic farming is still regarded as little more than a joke by the agricultural establishment. The National Farmers' Union dismisses it as an impossible prospect since, it argues, the move towards specialist farms — in other words, farms with either too little manure or not enough — has gone too far, with the country divided up between the livestock and dairy producers in the west and the arable producers in the east. And this attitude is shared by other established authorities. A plan to open a self-sufficient research farm in Charnwood Forest, Leicestershire, was opposed in 1979 by the local council on the grounds that "organic farming spoils attractive countryside".

There are about 250 organic farmers in Britain, about 0.1

percent of all farmholdings. Vast sums are spent each year on research and development into agrochemical uses and modern farming methods in an attempt to find a way out of the problem of spiralling costs that are eating their way into farmers' incomes. A small fraction of this, if it were to be spent on organic methods of farming, could make alternative husbandry even more efficient, in any logical or rational meaning of the word, than it is already.

A careful analysis of the farm of one of Britain's leading organic farmers, Sam Mayall, by the Department of Applied Economics at Cambridge University, found that its costs were below the national average, and its yields higher. Mayall and his son (Sam died in 1980) used no artificial fertiliser, though occasionally sprayed in an emergency — on docks, for instance. "We can find on this farm" (in Shropshire), Mayall said once, "anything scourging the country at the moment. But it's not damaging the cropping in the same way. Our principle is, don't wipe out pests and viruses but learn to get along with them. If you kill 75 percent of the pests by spray, you may find you've killed 100 percent of the predators. I'm not saying our crops are immune; just that they're more disease-resistant".

His grassland responded well to chemical-free treatment and his Ayrshire cattle continued to win prizes. Though at first wheat yields fell, they soon soared. The soil, a living organism, suffered initially from 'withdrawal symptoms' — as it has with many other farmers who have switched to organic methods — since it gets lazy after persistent applications of chemicals. (For example, a chemical-free field can produce ten times as much phosphate in the growing period as in the dormant period; the fluctuations are much less on sprayed fields.)

"All the time," concluded Mayall, "farmers are being urged to do the wrong things for economic reasons. Can we go on? Look how the Fens have dropped their soil — some are almost down to the subsoil".

A sceptical reader may be surprised at the number of studies that have been made. A report sponsored by the Dutch Agriculture Ministry in 1973 concluded that biological farms generally can produce a yield per acre comparable to orthodox farms, with the exception of dairying and market gardens. Farmers who deliberately chose to limit yields produced better-quality food.

Two years later, a report by the French Government concluded that biological farming did not necessarily reduce yields. In only two areas — maize growing and orchard fruit production — were yields lower than on conventional farms.

At the same time, the Cambridge University Agricultural Economics Unit carried out a survey of six organic farms, all of them mixed farms. It then compared yields to the performance of orthodox farms producing the same crops. It concluded that it was possible for organic farms to be as efficient as conventional ones in terms of yield per acre or per cow; cows on organic farms actually produced more milk.

An important study conducted over two years in the US corn belt concluded that organic farms fared better in bad weather, though in a good year, from the point of view of climate, output was a little lower than that of conventional farms. But the survey, by the Centre for the Biology of Natural Systems in the United States, found that labour costs were about equal, overall costs were lower on organic farms (which used about a third less fuel) and organic farms were just as profitable when all types of expenditure had been taken into account.

But these comparisons do not take into account the much higher water content in produce grown with the help of inorganic fertilisers, especially nitrogen. The weight of conventionally-grown food is thus much heavier. High water content also leads to problems of storing fruit and maintaining its quality.

I make no apology for returning to the problem — the crucially important problem — of waste. About half the amount of nitrogen fertiliser with which the land is drenched is not absorbed by plants. Instead, it drains away to pollute rivers, streams and lakes. While the Ministry of Agriculture says there is not enough manure for effective organic farming, millions of tonnes of slurry, much of it untreated, ends up somewhere in natural water courses.

Burning straw is wasteful, and also damaging to hedges and, when farmers are careless, to drivers passing by on motorways. Straw mixed with manure can be used in feed (and pig and cattle manure is a rich, cheap source of protein). But burning stubble kills off weeds, helps to kill diseases in the soil (encouraged by intensive farming practises). Above all, it saves labour, and, therefore, in the short term, money.

Soil is an extremely delicate, sensitive, and complex substance. We may regard it simply as a dark brown carpet, sometimes bare, sometimes covered with green, and occasionally hidden by a sea of waving, golden wheat. In fact, it is a living mass of micro-organisms and fibre and humus that performs vital functions, providing air, moisture, food and warmth. Tiny bacteria break down natural chemical compounds and live on vegetable and animal residues and fix nitrogen. Humus also absorbs toxic trace elements.[5]

Twelve years' investigations into the quality of crops grown in organic soils, compared to soils treated with chemical fertiliser, have shown an increase in dry matter of 23 percent, 18 percent of relative protein, 19 percent of sugars, 18 percent of potassium, 10 percent of calcium, and 13 percent of phosphates, according to the Soil Association.

To be consistent with the laws of nature, that is to promote the continuing diversity of species, protect the soil and recycle organic matter, the Association recommends simple guidelines for proper agricultural husbandry:

Rotate crops, it says; never bury manure and compost deeply; do not waste organic residues; pay particular attention to the aeration of the soil and the well-being of the earthworm population.

Research at the Government agricultural station at Rothamsted has demonstrated that natural manure is as effective as artificial fertiliser. Many would suggest that spelling out the dangers and damage inflicted by agrochemicals is merely 'crying wolf'. However, it should be remembered that as it takes a long time before illness or death can be put down to working conditions or exposure to certain chemicals, there is no immediate proof of damage to the soil. A one percent change in all the organic matter in the soil may take 40 years.[6]

Mixed, extensive, farming systems would not only be better for the soil. They would produce more jobs, and encourage a healthier diet. It would be less energy-intensive and less wasteful; animals consume between seven and ten times as many calories as they actually produce as meat. A more rational system would encourage more self-reliance. It would

The Principles of Organic Husbandry

The principles which govern the system are simple and the overriding one is to make use of natural resources and natural processes as far as is possible. This entails:

Returning to the soil all animal and vegetable residues. Waste nothing.

Handling manure so as to retain its full value by a system of rough composting.

Letting the soil feed the crops and avoiding direct feeding with soluble materials.

Avoiding the use of all chemicals liable to kill or lessen the activity of soil organisms.

Feeding the livestock as far as possible on the produce of the farm

The use of homegrown seed.

Practising mixed husbandry within the limits imposed by physical conditions and economics.

Minimal disturbance of the soil by cultivations, consistent with adequate weed control.

From: 'Farming Organically', The Soil Association, 1976.

help the root problems facing poor countries and challenge present trading systems. Three-quarters of US grain production is fed to animals and pets. Most of the protein imported by Britain is fed to animals. Livestock in the developed world consume about one-third of the world's cereals and a quarter of the fish that is caught. Developing countries, with 70 percent of the world's population, feed only 10 percent of its share of the world's cereals to animals.

We would not have to be vegetarians — far from it. It is simply a matter of regulating meat to a lower priority and feeding livestock on grass and pigs and poultry on more swill, and using animals more constructively. After all, the amount of nitrogen nutrient needed per acre of grass is available from manure produced each year by one cow or about ten pigs. According to some calculations Britain's organic waste is a food resource as valuable as half the country's cereal production.

We are wasting land as well as resources. By drenching some parts of the country with chemicals, we are ignoring other areas, notably the uplands. The productivity of hill land would benefit significantly from more investment in drainage and irrigation schemes, and at the same time keep pressure off valuable wetlands and marshes which are being threatened elsewhere. We could also spend the very little it would need to clean up our canals — an enterprise that would also provide jobs — to restore a network of unpolluted waterways.

There is plenty of land, even for much less intensive farming systems if we shift our diet away from animal products, something which even the Department of Health has suggested.

As one commentator has suggested:

"National food requirements based only on crop production could be provided from half of the present yield per acre — the yield level in the early part of this century when no artificial nitrogen was applied. In this way 500,000 tonnes of elemental nitrogen and 1,000 million litres of diesel oil would be saved. Clearly, an exercise such as this highlights the heavy price we pay, in terms of both resources in land and support energy, for a meat diet. Sooner rather than later, this price may be recognised as too high."[7]

Recommendations about our protein needs have been progressively reduced by nutritionists, by as much as 5 percent since the War. And we could switch to vegetables, notably beans, as a source of protein. But we need not all become vegetarians. Britain could support its population by a sensible agricultural system, not one based simply on the intense competition for profit.

Colin Tudge describes it admirably: "The structure of farming would vary from country to country. Most farms would tend to be mixed, because mixtures tend to make the best use of the land; yet terrain and climate would inevitably dictate that some areas were predominantly arable and some

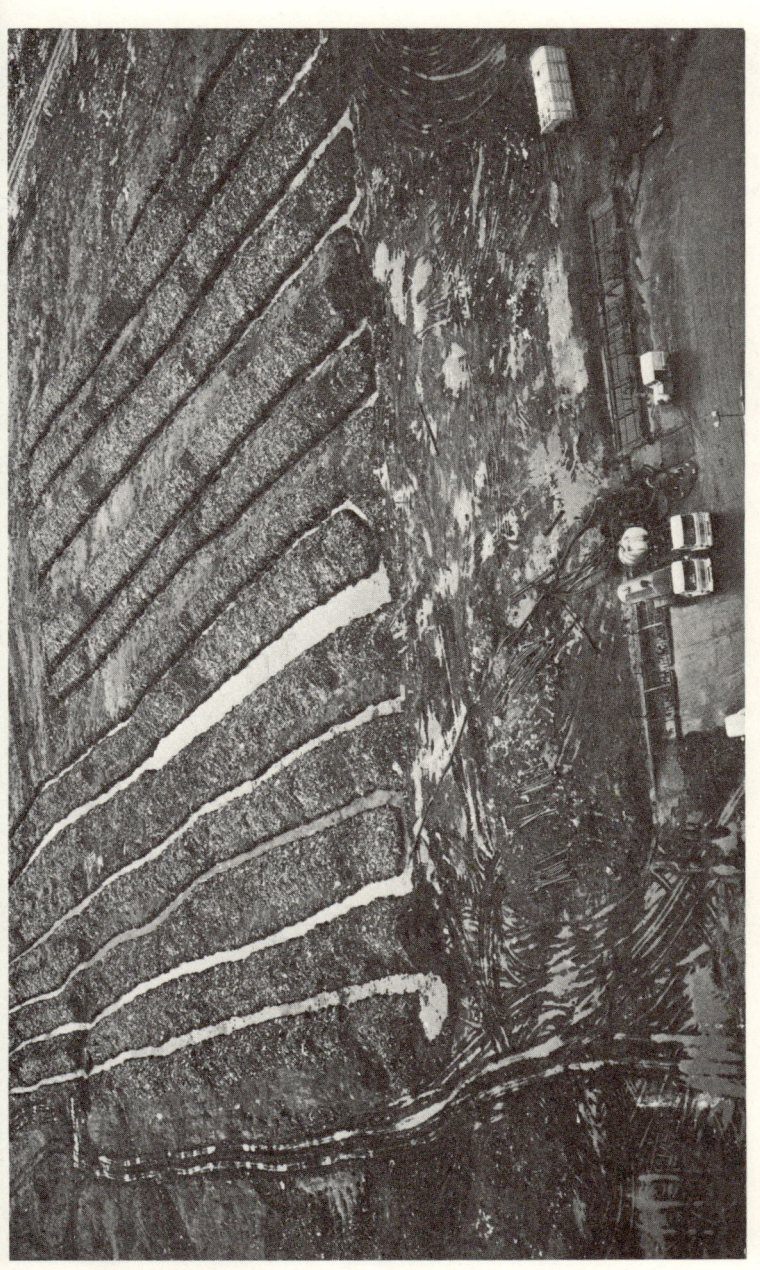

Pitsea Marshes. Whitehall, local authorities and private companies adopt a complacent and secretive attitude towards the dumping of toxic waste. Waste from as far afield as Cornwall and Scotland is transported to Pitsea, Essex, the country's largest toxic tip. Britain is also responsible for 90 percent of man-made radioactive waste known to enter the world's oceans. (Photo: Southend Air Photography)

pastoral . . . Rational agriculture cannot meet its full potential unless it employs a lot of people; not to take the place of the combine harvesters in some latter-day Luddite revolution, but to run the inevitably more intricate farms with their often elaborate interplay of crop and crop, and of crop and beast. . .

"So if Britain's agriculture were to become rational, it could also be relaxed. There would be no need to grub up yet more hedgerows in the interests of productivity. The old deciduous woods and copses could be left, and extended. The fens and the chalk grasslands, with their astonishingly rich ecosystems, could rest undisturbed. There would be plenty of room for people to live and enjoy themselves in the countryside; no excuse for those fences that now proscribe entry to the most remote hillsides, or for the tedious prairies that offer no way through."[8]

7

THE PLANNING MESS

"Our cities are a wilderness of spinning wheels instead of
palaces; yet the people have not clothes. We have
blackened every leaf of English greenwood with ashes,
and the people die of cold; our harbours are a forest of
merchant ships, and the people die of hunger." – *John
Ruskin.*

The recent history and experience of planning in Britain is a
supreme example of how good intentions have been thwarted,
halted and finally mocked by bureaucratic procedures and
short-term political considerations. Add to that a declining
economy, greed and speculation, and you have the sad state
into which the planning process has been dragged.

Green field sites have been plundered for new industrial
investment, yet there are at least 250,000 acres of dormant
land in urban and industrial heartlands; about 75,000 acres of
farmland are lost to urban and other uses each year —
equivalent to losing an area the size of Bedfordshire every four
years — yet official policy is that Britain should grow more
food; one million families are waiting for council houses, yet
one million houses are kept empty.

Well-meaning, but unimaginative, planners vie with impo-
verished local authorities and private property developers while
the land and the people suffer. Expensive office blocks are built,
building societies complain about a shortage of land, while
thirty square miles in London alone is derelict. Landowners
can freely scar the countryside, roads can tear through prime
farmland, while urban householders spend months nego-
tiating just to change the porch.

"We act", says Dr Alice Coleman, geographer at London

University, director of the second Land Utilization Survey and one of the few who recognise there is something to worry about, "as though we lived in the centre of an infinite virgin continent instead of a relatively small island".

The lack of concern, the complacency about what we are allowing to be done to the land, and about those who are doing it, is reflected, once again, in our ignorance. No reliable figures are kept on land use, on how farmland is lost or switched to different types of farming or forestry, on vacant or on derelict land. It is not that the Government is not willing to tell us; it is simply that Whitehall itself does not know. So we are alerted only to particular scandals or particular disputes; the destruction of a beauty-spot, for instance, or a decision to concrete over urban parkland — and then the controversy dies away. These are symptoms: the cause is our neglect but the neglect, above all, of those who are in the position, or have the money, to take decisions.

Local authorities were asked by the Department of Environment in 1974 for statistics on land use. Three years later, only one authority in eight had bothered to reply. Perhaps indolence was not the sole reason; seventy-five percent of vacant land in cities is in the hands of public authorities.

The Civic Trust conducted a survey on urban waste land in 1977 after the Department of Environment admitted that it had no reliable figures on idle land and insisted that a proper study could be provided only at 'disproportionate cost'. Dr Coleman's second Land Utilization Survey, which began in 1963, was funded by Sainsbury, the supermarket chain.

"Precise information on land transfers between different uses is not available", says Whitehall. But the Centre for Agricultural Strategy at Reading has warned that a scarcity of land is very likely by the year 2000. It also estimates[1] that 80 percent of urban land requirements in the next twenty-five years could be met from existing waste land. But the Department of Environment merely points out that for the rest of the century "losses of agricultural land will continue to be more than offset by increased productivity". We have considered above what damage this presumption about greater productivity has already inflicted on the land and how it has led to more costly, capital and energy-intensive farming methods and reliance on chemicals. In addition, reliance on greater productivity will increase pressure on the landscape and wildlife and prevent people from enjoying the country. It would turn the countryside into a food factory.

The post-war history of planning in Britain offers ample ammunition to those, whether farmers, property developers or politicians, who argue that what are needed are fewer controls, not more. While large financial institutions and pension fund managers have enjoyed increasing financial power, planning bureaucrats have tried to determine how our land should be used. In Britain, over the past fifty years the questions how people should live, where they live, the whole shape of their environment, even where they have to do their shopping, is an awful testimony of how bureaucrats have secured a degree of physical control while the political environment and financial system remain untouched. They have succeeded in tearing communities apart, wasting both human and physical resources in the process.

There must be something seriously wrong with our priorities, with a system in which controls and endless correspondence with local authority planners are needed before the owner of a house in a town can make small alterations to a roof or porch, but which allows historic buildings, officially preserved, to be neglected and pulled down, prime farmland to be torn up, and landowners to scar the countryside.

The Town and Country Planning Act of 1947 was a piece of pioneering legislation, widely welcomed as a progressive and enlightened initiative. It promoted the concept of zones of development, 'planning permission', and it gave to the State the authority of compulsory purchase. Since the Second World War, no one has been allowed, officially, to develop or redevelop urban or industrial property and land without prior permission.

One of the problems was that intervention by the State increased just at the time the public sector was being deprived of the necessary financial resources. If the State had the authority, it did not have the means to do much with it, while the private sector at least had the motivation of making money. A 1975 survey by the Department of Environment of

South-east England showed that half the residential land available for development was being freed as a result of appeals to Whitehall by developers against decisions (basically, to do nothing) by local authorities. We should not forget that most derelict land is in the hands of local councils.

The 1975 Community Land Act and the Development Land Tax were designed to give local authorities more power and the means to control the activities of private landowners and speculators. The so-called Community Land Scheme, promoted then by the Labour administration, was designed to "enable the community to control the development of land in accordance with its needs and priorities" and "to restore to the community the increase in value of land arising from its efforts".

What happened? The 80 percent Development Land Tax discouraged private developers from putting land on the market. They preferred either to keep hold of their land or pass on the tax, thereby increasing the price of land. More significantly, local authorities had to rely on land funds provided by central government. Public expenditure cuts quickly took their toll and in 1976 the Treasury halved the amount of money made available for the purchase of land. Local government plans were subjected to detailed scrutiny by civil servants in Whitehall. In the first year of the Community Land Act — which was introduced amid anguished cries of 'creeping socialism' — local authorities in England bought just 1,571 acres of land at a cost of £12 millions. In the second year they bought about 800 acres. Some 70 acres were bought in Scotland and 1,000 acres in Wales (where the Development Agency adopted a much more coherent and positive attitude towards the scheme).

Yet this paltry achievement by 1978 had involved nine orders and regulations made jointly by the Secretaries of State for the Environment in Scotland and Wales. There were, in addition, four English orders, 23 English directions, 35 English circulars or letters, 6 Scottish orders, 2 Scottish directions, 34 Scottish circulars or letters, 2 Scottish publicity booklets, 13 Welsh circulars, 5 Welsh guidance notes and 2 Welsh directions to the Land Authority.

Delays and uncertainty helped to fuel speculation, increase

costs, and force up prices. Either because they were short of money or because they deliberately hoarded land in the hope that they would get a higher price, or more attractive offer — from office block developers, for instance — local authories were no more willing to release land than were private landowners.

The problems were compounded because the Community Land Act specified that land bought for redevelopment had to be purchased by local authorities at a price based on the value of the land to the seller rather than the buyer. Thus, the London borough of Camden, for example, had to pay £175,000 an acre for 43 acres of obsolete railway land at a price reflecting some previous use value — or potential development value — to British Rail, rather than any current or specifically housing value to the council.

Graham Moss, an architect and member of the Land Decade Educational Council, summed up the problem this way: "The Community Land Act, introduced in 1975 in response to massive land speculation by developers in the 1960s and early 1970s, emerged as law in a period of economic recession parallelled only by the 1930s. Together with Development Land Tax, both pieces of legislation began operating at a time when scores of developers were declared bankrupt, and during which many local authorities were only too thankful to see private development taking place. The final irony came when, in 1979, the Community Land Act was withdrawn, just at the very period of development activity for which it had been originally designed was returning".

An example of how the lack of 'public money' has enabled private companies to take over the property market, in spite of controls, is the Centre Point tower, built on council land in London. The Oldham Estate Company bought much of St Giles Circus, wanted by the then London County Council, for an ambitious traffic scheme, at a price the council could not afford. It donated it to the Council. In exchange, the company was given preferential planning permission. As the Cambridge department of Land Economy put it: "the Oldham Estate Company was in a position, as a single landowner, to manipulate market forces, in this case the supply of a particular tract of land, to bring about development and to

benefit from the increase in land value. The subsequent
development of Centre Point, over a traffic development that
never actually came into being, remains a controversial issue".
The building was occupied by squatters, left empty and
became a symbol of waste and white elephants until it was
finally occupied by the Confederation of British Industry
in 1980.

Public authorities have to compete with the private sector at
the going rate on the open market, with the additional burden
of having to use most of the money in their housing account to
pay interest on loans. Not only do they have to subsidise
council housing — as much is spent on this as is spent on
subsidising mortgages for private house-owners — they spend
about two-thirds of their housing budget on interest on
borrowed money. It is therefore scarcely surprising that in
1977 Islington concil in London reversed an earlier plan to use
derelict and vacant land for housing, shops, and light industry,
in favour of an £8 millions office development scheme.
"Islington's greatest need," said David Hyams, chairman of
the council's planning applications committee, "is new jobs.
People working in the Angel will bring life back to this blighted
area".

But what happens when a council wants to use land for
homes? Lambeth was the first London council to take
advantage of the provision in the Town and Country Planning
Acts to draw up a district plan. It wanted to build homes, but
the Greater London Council, then Conservative-controlled,
pointed out that this would not bring in money. It urged Peter
Shore, the Labour Environment Secretary, to persuade the
borough to change its mind and refused to sell the land to
Lambeth at its historic, cheap, price if the land was used "only
for homes". The GLC planned to build a 600-bedroom hotel
on the six-acre site on the South Bank adjoining the National
Theatre, complete with a 380-foot tower, 48 luxury flats and
offices, though as a belated gesture it did agree to add 200
council houses. The complex was estimated to generate £2
millions in rates, against the £70,000 the homes-only plan
would have provided in income to the council. Local community associations pointed out that the proposed hotel
would be just a few hundred yards from the existing Melia-

Buckley hotel at King's Reach which had not even opened though applications had already been made to convert its surplus bedrooms into offices.

Housing

Though apparently short of money to fulfil the needs of their most deprived sections of the community, local authorities waste millions of pounds 'modernising' houses, buying up property, and leaving them empty. It is partly the result of an inflexible budgeting sytem, whereby a certain amount of cash has been earmarked for a particular purpose, and cannot be switched to another, more important, part of the budget. Once staff have been assigned to a parks division, for instance, they will not be transferred to the housing department. It is also the result of a bureaucratic culture which equates the size of a budget with influence and potential power and, in this case, regards the accumulation of property as more significant and important and as an end in itself rather than a means of housing more people.

In 1981, seven local authorities — all Labour-controlled — each owned more than 1,000 homes which had been empty for more than a year. Top of the list was Manchester, but five were in London: Islington, Hackney, Southwark, Lambeth, and Camden. The councils blamed central government for not giving them enough money.

There are about 1 million families waiting for council houses in Britain, and about 1 million houses, with both private and public landlords, are kept empty. Over 3 million people live in overcrowded homes. The Building Research Establishment estimated in 1979 that 3 million homes in England and Wales were unfit to live in or lacked basic amenities. "The number of houses declining into unfitness or major disrepair," according to Shelter, "is now equal to the number of houses being improved and slums demolished".

The number of homeless families was estimated in 1978 at 52,000 — double the 1971 figure and four times the number in 1966. It is increasing. London's wasteland, of 25,000 acres in 1980, was enough to house the city's homeless while the only remaining countryside area in East London — Walthamstow Marshes — was at risk because the Lea Valley Regional Park Authority wanted to quarry it for gravel.

At the beginning of 1978: fewer houses were being built than at any time over the past twenty-five years; 300,000 building workers were unemployed; about 8,000 architects were jobless; gazumping was on the increase.

In the five years up to 1975, owner-occupiers spent about £1,250 millions in fees to estate agents, solicitors and surveyors — enough to build more than 155,000 homes. There were at least half a million more houses than households, yet twice as many houses kept empty (there are no official figures on the numbers of houses kept boarded up).

There were an estimated 350,000 second homes in Britain in 1978 and the number is growing by 25,000 a year. 1 in 3 of all houses in the Lake District are second homes, empty for most of the year.

In the five years up to 1979, the Government actually reduced the amount of money spent on building new homes and buying or repairing old ones by 40 percent. Housing's share in total public spending was cut. The problem is getting worse.

The figures illustrate the paradoxes and absurdity, let alone the scandal, surrounding the whole question of housing. Freedom to have a home is one of our basic birthrights, yet it is treated with disdain, even frivolity. As land is allocated according to profit, so housing is always at the back of the queue. It is well put by the Community Development Project*: "as soon as the poorer sections of the working class have access to enough money to house themselves adequately at current prices, the market process tends to push house prices up beyond their reach. It is built into the existing system that it is never possible for everyone to buy a decent place to live".

So in addition to the planning scandal, there is the financial scandal of housing policy in Britain, including an artificial

---

*The CDP was set up by the Home Office in 1969 as a "neighbourhood-based experiment aimed at finding new ways to meet the needs of people living in areas of high social deprivation". But its reports became increasingly controversial, and the Home Office closed down the CDP seven years later.

housing shortage which has tended to raise prices well above inflation rates. And rents for council homes have increased substantially over the past few years at a time when over a hundred local authorities pay out more in interest than they get in rents. The way hard-pressed councils pay more than they should for land was well illustrated in the Bewbush land affair. In 1976, Bewbush Manor in Sussex was sold for £3.25 millions. Three months later, half the land was re-sold to Crawley council for housing at a price of £7 millions, with the estate agents involved taking their usual commission as well as a share of the profits.

Money is being spent, land and houses wasted, to preserve an extremely inefficient, divisive policy. (The Conservative Government's policy to sell council houses, even with large discounts, will not solve the problem since the poorest and most needy will be unable to use the scheme, and those who live in some of the worst tower blocks in an unattractive environment will not want to purchase flats there.) There is no greater and more obvious manifestation of 'class barriers', however disguised and artificial it has generally become, than the division between those living in their own private accommodation and those living in council property — the latter in theory looked after by public authorities but in practice at the mercy of the monopoly tendencies of the State.

Public authorities have achieved no more for housing than private developers have achieved in the commercial land market. Tower blocks were the answer of planners, encouraged enthusiastically by pundits and the media, for people displaced as a result of slum clearance programmes. It would have been better in many areas simply to have renovated back-to-backs, giving people more space, a garden of their own, and, above all, preserving existing communities. The people making up those communities were never asked what *they* wanted.

The social and environmental consequences of tower block estates are now well-known. To be specific: the Ronan Point explosion occurred in 1968, yet it was not until 1974 that the Greater London Council decided not to house families with children or old people in buildings of more than four storeys. Would the private sector, had it been responsible for all

Ronan Point. *The Ronan Point disaster in East London, 1968, focussed attention on tower blocks and modern construction techniques. In three years in the early 1960s, 110 tower blocks were built in Liverpool alone. (Photo: Assoc. Press)*

housing, have stopped much earlier building accommodation which people manifestly did not like?

The answer must lie in an environment which allows people to decide, in the framework of smaller communities, what housing system they want. By definition, this means some degree of flexibility, abandoning dogma, state legislation and interference imposed by the centre. It surely is one of the greatest and astounding ironies that the turnout in local elections is so low, typically around 30 percent, and that so much media interest is attached to the House of Commons and so little to secretive local councils which have so much control over our education services, our environment and over housing policies.

All sorts of practical measures can be taken. Liberal councillors in Liverpool, for instance, have built attractive groups of houses, offering them to the tenants at prices well below the prevailing prices on the open market. Rents could contribute to the purchase price of the tenants' houses; tenants could own their houses after a certain time. There are plenty of different possible schemes and arrangements that would cut through the present divide between landlords and tenants on one side, and owner-occupiers and building societies on the other. Values put on houses or flats should be expressed in terms of need — or simply what is appropriate or common sense — rather than income. The division between 'poor' or 'bad' areas and 'good' or 'polluted' areas should be ironed out. The priorities of both local and central government must be changed.

The Waste and Loss of Land

With her students, Alice Coleman, one of the few individuals in Britain who are angry about the disastrous history of planning in post-war Britain, surveyed every metre of the London borough of Tower Hamlets in the summer of 1977. There she found 884 abandoned buildings, 45 kilometres of corrugated hoardings, tower blocks with 567 dwellings accessible from one single entrance, a high incidence of vandalism, a massive increase in derelict land over the previous ten years, and a 134 percent rise in planned 'open space'. As Coleman put it, the greater the amount of derelict land, the higher the rents and rates on the houses that remain.

One-third of the borough was 'open space' of one kind or another: on one council estate, only about a quarter of the land consisted of buildings, and more than half was open space — the rest was used for garages and sheds. The point is that no-one uses the open space; indeed, they are not allowed to. No walking on the grass, no ball games here. About three-quarters of Tower Hamlets was publicly-owned in 1976, and the public authorities treat adults like children, and children like babies.

We now know the problem. The council does not build homes because it has insufficient funds. It hoards the land. Private developers are not prepared to pay high prices for land (prices in some cases are raised artificially because of controls). In any case, they are not prepared to buy land for low-cost housing for low-income families. The council, too, is after office-blocks or luxury dwellings from which they can get high rates. Those who are not in the tower blocks or council estates are uprooted, or give up and drift away.

Official acreages of Derelict Land in England & Wales

year	reclaimed	end year balance
1964	2076	84,900
1966	1641	92,876
1968	2113	93,920
1970	3645	96,697
1972	5360	n.a.
1974	6978	106,918
1976	n.a.	137,000

Note: Derelict land, which is mainly an urban phenomenon is distinct from Waste land and Scrub, which are mostly rural, and for which the acreages are much higher than for Derelict.
(Source: Land Decade Educational Council)

Whitehall officially estimated that there were 137,000 acres of derelict land in England and Wales in 1977. But there are no official figures on dormant land (undamaged land lying idle) — the sites, for instance, of demolished buildings or abandoned allotments. Much is in temporary use, notably as car parks, but is crying out to be acquired for more useful purposes.

The 1977 Civic Trust survey showed that there were about 250,000 acres of such dormant land — equivalent to the cities of Birmingham, Derby, Glasgow, Hull, Liverpool, Manchester, Nottingham, Portsmouth and Southampton combined. That is sufficient to provide housing for 5 million people. Some sites, the Trust acknowledged, would be difficult and expensive to exploit, but most of the land lying vacant, wasted, is well located and simply waiting to be used.

There were 800 acres in Southwark in 1976, 80 percent of which was owned by the council, including the Surrey docks. Greater London had over 20,000 acres of vacant land and buildings, a tenth of it in the inner city — an amount almost equal to the entire designated area of Hatfield new town with a population of 26,000.

The area of derelict land has increased by 40 percent over the past twelve years. Wasteland in the Thames area nearly doubled between 1962 and 1972; but in the past five years, the area of vacant land in inner London has also doubled. In 1978, there were more than 10,000 acres of derelict and wasteland in the West Midlands, over 20,000 acres in South Yorkshire, and 9,400 acres of neglected, idle, land on Merseyside.

The Civic Trust conducted a detailed survey of 279 vacant sites throughout the country. It showed that 66 had been homes, 22 railway stations and railway land, 17 quarries or tips, 15 had been shops, 12 gardens, 10 churches or chapels, and 7 allotments. Local authorities owned 39 percent of the land, private companies 13 percent, and 11 percent were currently in the hands of builders and developers.

In the same week the Civic Trust report on derelict land was published, the sale was announced of about 15 acres of farmland in the Severn valley, on the edge of the Gloucester-shire village of Highnam. Planning permission had been given for the construction of 123 houses. A scheme to make use of wasteland around Gloucester docks was quietly dropped by the British Waterways Board.

Edinburgh is still an attractive city, but it can also be proud of more than 100 derelict sites acquired by the local regional and district councils for road-building schemes, nearly all of which have since been abandoned. The University wanted to demolish historic Georgian buildings in George Square though a site for more than 100 houses around the corner in Charles Street remained undeveloped.

Worcester district council demolished a terrace of early eighteenth century houses, and a shop, on 4 acres bought in the 1950s for road widening. The land remains unused.

The inhabitants of Yarm, Cleveland, abandoned their hopes to use British Rail land for allotments because of the excessive rent demanded for a short-term lease.

During the Beeching period (1965-70), British Rail land equivalent to a conventional town of 80,000 people was released each year. BR (whose chairman, Sir Peter Parker, is a trustee of the Civic Trust) is under a statutory obligation to act 'commercially', yet incurs no penalty for neglected land. While railway land was being made idle, over 2,500 acres of mostly agricultural land were being eaten up for motorways and trunk roads.

The railways and waterways of Britain are neglected, yet Britain has the unenviable distinction of having more miles of asphalt road compared to total land area than any other country.

In Witham, Essex, the local countryside society suggested that 25 acres owned by Bovis, the construction firm, should be farmed pending development. It found a farmer willing to go ahead, but the company blocked the idea. "We are afraid," a company spokesman said, "that the tenant might acquire a right to permanent occupation".

Brighton Civic Society unsuccessfully defended a school widely regarded as a unique example of Regency Gothic architecture. It was demolished in 1971 to make way for a road widening scheme in spite of the well-publicised intention to list the building. The road scheme fell through and a pedestrian precinct was put in its place.

In January 1977, the London borough of Lambeth had to

Decayed Lock near Winnock. *Most of the old waterways of Britain have been neglected, while food accounts for more commercial road traffic than any other single item in the economy. (Photo: Denis Thorpe)*

bring in a line of policemen to protect workers demolishing homes in St Agnes Place which were to be cleared to enlarge Kennington Park, even though the council admitted that it would not have enough money to carry through the scheme until well into the 1980s. (A High Court injunction stopped the demolition and the council later reversed its decision.)

A month later, in February 1977, Peter Shore, the Environment Secretary, said: "I think we do now realise and accept that the decline of the inner cities has gone too far and too fast". During the past fifteen years, the inner areas of Manchester had lost 20 percent of their population, and of Liverpool no less than 40 percent. The Home Office urban deprivation unit described, "the bleakest planned bomb sit in Britain stretches for 3,500 acres along the River Clyde — the wasteland of the east end of Glasgow, from Glasgow Green east to Shettleston, south to Dalmarnock and Auchenshuggle. This was once the powerhouse of Clydeside's heavy industry.. Along the London Road or Tollcross Road, arteries through the areas, facades of a late Victorian city shield, literally, miles of dereliction".

90,000 houses a year have been pulled down in inner cities in the name of 'slum clearance'. Simultaneously, planners have encouraged, even forced, people to live and invest in new towns that are geometrically organised, cold and claustrophobic. And in between the cities and the new dormitory towns, you find miles and miles of sprawling urban fringe. Suburbs mix with sewage works, reservoirs, rubbish tips, golf courses, caravan sites, a landscape crossed by pylons and a dense network of roads. The distinction between town and country is blurred. The Civic Trust found that on 69 farms between London and St Albans, nearly 90 percent were affected by trespass, and over 70 percent by the dumping of rubbish, and many by cattle and sheep rustling, and theft.

8

THE MESS ON THE FRINGES

It has been estimated by members of the Land Decade Educational Council that at the end of the 1970s, there was a total of 1.3 million acres of derelict and idle land, plus an additional 1.2 million acres at risk from industrial pollution or affected by special access rights, planning and other legal restrictions. This is an area that amounts to about 4.5 percent of Britain's land surface; twice the size of Northumberland.

I have, over the past few pages, listed many examples, and provided many figures, to illustrate what is happening, to paint a picture. There is probably an example around the corner from where the reader lives, of the waste, neglect, complacency and irresponsibility that permeates any discussion of how we have treated our land. All this has happened in spite of planning laws and can be seen perhaps most dramatically in that grey area, which includes 'the green belt', astride the towns and countryside. There is land there which is the victim of such neglect that nobody can use it, or wants to use it, and therefore prey to such schemes as the M25 outer circle orbital London motorway that will attract more urban sprawl and industrial estates. Coleman has estimated by studying her land use maps that between farmland on the outer edges of the urban fringe and our 'inner city deserts', there is enough land for all development needs for the next 50 years. Development, she recommends, should be banned anywhere else.

Farmland in Surrey has been seriously fragmented by roads, rubbish tips and speculators buying property and deliberately allowing it to become derelict to make it easier for planning permission for housing or industrial development. Land is used simply to graze horses. Land is sold off in small plots by hard-pressed farmers to developers, for golf courses,

Interchange of M22 and M25: *This maze south of London covers 113 acres of good agricultural land. Each mile of motorway consumes up to 40 acres of land and some 10,000 tonnes of gravel etc mined from elsewhere. (Photo: Cyril Bernard © Camera Press)*

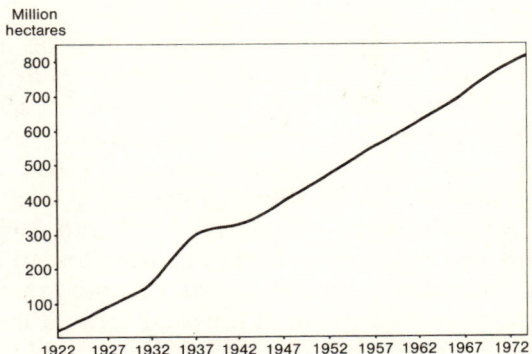

24. Loss of Farmland: *The graph shows the cumulative loss of farmland. Net losses to all other uses, including forestry, amounted to 148,000 acres a year in the mid 1970s (ref. E.D.C. for Agric.) Some argue that Britain is producing more food from less land, through more intensive farming methods. Others (eg Alice Coleman) say the increase notably of cereals, has been achieved from more land, as pastures are ploughed up for arable crops. (Source: Alice Coleman)*

or industrial firms. Some of the biggest farms in the area have been sliced by the M3 and M25 motorways (the latter consumes 25 acres of farmland per mile). Gravel extraction has destroyed large areas of the Thames valley. Though Surrey County council do not have reliable figures, an estimated 3,750 acres in the Runneymede area has been excavated for gravel, and none restored to good quality farmland: they are either turned into reservoirs and water parks or are left to their own resources and become poor grassland.

The map on page (?) shows that most of the best farmland adjoins urban centres. Much of it has been taken over already by urban or industrial development. At the present rate of dereliction and development, all the farmland in the area comprising the Thames estuary, Surrey and Merseyside will have vanished within 85 years.

Land Lost to Agriculture

Between 1955 and 1975, over 2 million acres — 100,000 acres a year — were lost to forestry, urban and industrial uses.

Between 1971 and 1978, 125,000 acres were lost annually, but this excludes rough grazings.

The Ministry of Agriculture assumes that up to 1983, 104,000 acres will be lost to food production each year.

More than half of these totals are accounted for by forestry and woodland; the rest to urban and industrial development. (The loss to urban uses was greatest in the 1930s.)

Alice Coleman has shown that between the first and second Land Utilisation Surveys — between 1933 and 1963 — 1.4 million acres of good quality farmland were lost in England and Wales. She also points out that upland grazing land accounts for about 15 percent of the total agricultural area in England and Wales but for only about 2 percent of total food production.

In a report on land use in 1977, the Agriculture EDC warned: "If the present economic conditions persist, authorities are likely to be more than usually attracted to the land which is least costly to develop, i.e. farmland of higher quality. It also noted that the Department of Agriculture and Fisheries for Scotland forecast that half the land for development will be taken from Grade 1, 2 and the upper range of Grade 3 farmland of which there is a particularly small proportion in that country. (See fig 8, p57)

There is an argument, which seems to be gaining currency, that there is plenty of land about, since agriculture is becoming increasingly productive. But this evades the question of what kind of agricultural system we want, as well as the amount of waste land in inner cities. Despite attempts to attract residents and businesses to inner city areas, green field sites continue to be attractive as easier and cheaper locations for builders and developers.

The Loss of Farmland

There will still be plains of grain in East Anglia, small pastures in the West, and in hill areas. But while the remaining farmland in the lowlands will be under tremendous pressure to produce more food with intensive farming methods, hill area will be carefully organised as the last remaining outposts for organised recreation for 'the urban masses', huddled and squeezed together in a wilderness of dereliction.

> The National Economic Development Council estimated in 1977 that 148,000 acres a year were lost to agriculture: 100,000 acres to forestry, and the rest to urban uses, most of which were not subject to planning inquiries and made up, as one observer put it, of "the residue of the urban fringe".
>
> But according to the latest Ministry of Agriculture figures, 78,400 acres were lost to farming in 1974-75: 32,000 acres to urban uses, 3,100 acres to forestry and 44,100 acres to what is called 'other adjustments'.
>
> The Ministry, however, acknowledges that it has no reliable mechanism or yardstick now to measure the shifts from one land use to another.

Farmland has been taken away from agriculture to urban industrial and road transport development at an alarming rate just at a time when urban dwellings have become more, rather than less, intensive. The average annual loss of farmland, about 40,000 acres, or more than double that if forestry is included, between 1969 and 1974 was thirty-seven percent greater than between 1955 and 1960. Between 1970 and 1975 no less than 500,000 acres was lost to farmland, an area the size of Leicestershire.

Over the past few years, 153,000 acres of farmland were lost in North-West England while the amount of derelict and vacant land in the region increased by 23,700 acres.

We have noted that Grade 1 land has been particularly vulnerable to development. In the course of her map-making, Coleman discovered that Grade 1 land had been built on five times as fast as poorer quality land. Yet the product of prime land can be six times as much as rough pasture. It is often the best land which is ploughed up to be used for just a couple of year's supply of gravel.*

How is this consistent with the Government's stated object-

*The Ministry of Agriculture itself acknowledges that there is a real danger that the peats of the Grade 1 land in Lancashire and the fens will be blown away in forty years.

ive to become more self-reliant in food and to increase agricultural production? The loss of farmland, whether through industrial or urban development, roads or mineral workings, encourages farmers to adopt ever more intensive, and therefore costly, methods, to plough up moorland, dig up hedges, drain marshes, and threaten the countryside and wildlife.

Farmers rightly complain about the loss of farmland. But the losses also give them an excuse to plough up more land, invest in ever more intensive agricultural methods, switch from traditional grass-based livestock farming or mixed farming, to high-yielding and remunerative specialist cereal crops,. and demand even higher prices for their products. Above all, by being able easily to indict planning legislation, they insist that agriculture should remain exempt from it. We shall consider some of the broader implications of this in the following chapter. Is there a prima facie case to extend planning controls to agriculture? Yes. It all depends on what the controls are, their criteria and *who* decides. Farming should not be left to farmers any more than war should be left to generals. (And while there is an increasingly healthy public debate about defence policy, there is little or no debate about food production.) Much greater thought and enlightened planning would also produce a more coherent agricultural policy — something which, as we have seen, is absent in Britain today.

"It is no defence of agriculture's exclusion from planning to assert that farmers can understand farming decisions; we have long accepted that planners can decide upon other industrial and commercial processes of which they have no intimate understanding", according to the British Association of Nature Conservationists. It adds: "we would argue that a decision to improve a tract of open moorland or to drain a water meadow is not inherently more intractable than a decision to build, say, a hypermarket or a housing estate".

An immediately obvious candidate for controls is intensive livestock farming, recognised by the Royal Commission on Environmental Pollution as more akin to industry than agriculture. The Humberside County Council, in whose area are reared 10 percent of the country's pigs, broke new ground in 1979 when it drew up an Intensive Livestock Units Subject Plan. Not surprisingly its proposals were initially opposed by local farmers. But the Plan was the result of constructive debate between the council environmental health authorities, the Yorkshire Water Authority, the National Farmers' Union, the Council for the Protection of Rural England and the Ministry of Agriculture. Its main aims are to avoid pollution, improve existing units, encourage straw-bedding systems, ensure that there is sufficient land for spreading manure, that there is adequate storage capacity for manure, and that new housing is located away from existing units and that no new units are built in existing housing areas. When the Plan was being prepared those involved found that no scientific analysis of the potential problems existed: but a consensus emerged after reasonable discussion between local interests and on the basis of common sense. Some problems remain, but at least they are being discussed.

Since the production of food, and the kind of food produced, is so vital to each one of us and to our health, our income and family budgets, the matter must not be left up to 200,000 farmers. They have their own vested interests, after all, and they do not necessarily reflect those of the rest of

Land use changes in England and Wales

Category of use	1933 (acres)	1963 (acres)	change (acres)	percentage change (category)	(national area)
tended open space	266 582	461 834	195 252	+ 73	+ 1
airfields	30 755	108 478	77 723	+253	–
other settlement	2 399 235	3 452 263	1 053 028	+ 44	+ 3
total settlement	2 696 572	4 022 575	1 326 003	+ 49	+ 4
permanent pasture	16 463 606	10 919 028	−5 544 578	− 34	−15
leys	2 243 345	4 784 556	2 541 211	+113	+ 7
arable land	7 352 671	9 116 909	1 764 238	+ 24	+ 5
orchards	259 719	274 283	14 564	+ 6	–
allotments	60 812	48 849	– 11 963	− 20	–
total improved farmland	26 380 153	25 143 625	−1 236 528	− 5	− 3
woodland	2 115 464	2 993 889	878 425	+ 42	− 2
other cover types	5 825 761	4 787 850	−1 037 911	− 18	− 3
water	321 359	394 375	73 016	+ 23	–
total cover types	8 262 584	8 176 114	– 86 470	− 1	–
total area	37 339 309	37 342 314	+ 3 005		

(Source: Alice Coleman & The Architects' Jour.)

society. An informed debate about agricultural policy, and its planning and long-term aims, is all the more crucial since Britain will no longer be able to rely on its traditional role as an important trading nation and one of the world's largest food importers. Other countries have their own mouths to feed and political upheavals in many parts of the world, notably in developing countries, are likely both to disrupt trade and force Governments to pay more attention to feeding their own people. And, after all, it was the partial blockade of Britain during the Second World War which encouraged the original Town and Country Planning Act of 1947 and gave it widespread support. The aim — to increase food production and provide security for farmers — also lay behind the 1947 Agriculture Act, the basis of Britain's farm grant and price support system until Common Market entry in 1973.

Alice Coleman

Anyone interested in the land, and its neglect, owes a debt to Coleman and her valuable, dogged, pioneering work, first described for a broader, though still restricted, public, in an essay in the *Architects' Journal* (19th January 1977).

Pitching straight into the problem of pre-war misuse of land, she said: "urban fringe resulted from the 'dream of spacious living', which had seized the imagination of the interwar population. Released by transport developments from the necessity of living near their work, people had swarmed into the countryside in order to enjoy more spacious and aesthetic rural surroundings, while continuing to commute to town for more lucrative urban jobs".

Unfortunately, however, "the dream of spacious living is not sustainable on a mass scale", she argues. "The first-comers saw their peaceful spaciousness become congested with like-minded commuters and their vistas of rural beauty degenerate into weedy wasteland which often attracted rubbish dumping. The second wave of dreamers rejected this dreary prospect and moved on into the next zone of unspoilt farmscape, there to initiate the same wasteful and destructive procedure".

Before we go any further, we should understand Coleman's

main objective; a strict system of land classification. She has drawn up five main existing categories of land: townscape — predominantly urban settlements, such as inner city areas; farmscape — unspoilt farmland, the plains of East Anglia, for instance; and wildscape — poor land in hilly and mountainous area, for example, much of which could be put to better use, with forestry a prime candidate. The two other categories, what she calls 'rurban fringe' — the areas on the edge of large urgan conglomerations, and 'marginal fringe' — farmland which has been allowed to deteriorate in both hill and lowland areas — should, she says, be progressively turned into one or other of the other distinct categories.

Coleman drew up the second Land Utilisation Survey in the 1960s with the help of volunteers. They made 6,500 maps, 6 inches to the mile, showing about 250 different kinds of land use. They are available to the public at King's College, London University.

Then she took a second sample survey, comparing the state of the land in 1972 and 1962. It made some dramatic discoveries. The Thames Estuary area gained 15 times as much extra open space as extra housing (many older houses having been demolished). The net gain in land for transport, mainly roads, was 16 times as great as the net gain for housing, and the gain in roads was 22 times as great. (If we continue at that rate the whole of England and Wales will be covered by roads in 670 years.) The growth in wasteland was 61 times as great as the net gain in land used for housing and there was an increase of no less than 90 percent in the amount of derelict and idle land.

Coleman came to the conclusion that one of the main problems was the growth of the planning bureaucracy and the reverence accorded to it. Planners have no vision, they cannot see the woods from the trees, they became too specialist. They could take no broad view, they were unaware of any pattern of land use. For example, when agricultural productivity is discussed, land wasted is ignored.

Farmscape: tracts of agricultural land. It is being lost at an alarming rate, while good farmland that remains is under threat from indiscriminate use of chemicals. As we have seen, the Ministry of Agriculture has underestimated the amount of good land lost, partly because it does not have any identifiable figures for land left idle.

Urban fringe: in almost all counties, the area of urban fringe amounted to about twice the area of townscape. It is an example, according to Coleman, of how "land use planning has completely failed to achieve one of its chief objectives". It has been allowed to develop in Lancashire and Cornwall in particular, and in Berkshire where the spread of housing development has taken twice as much farmscape as was eventually used for houses and their gardens. Wasteland has grown most dramatically in the Thames Estuary area.

There is a use for 'functional fringe', areas serving urban settlements — yet large tracts of wasteland has been left between gasworks, for instance, abattoirs and other public utilities. There has been no economical approach about how to earmark certain areas for different functions, or groups of functions.

In the twelve years, 1964-1976, the area of derelict land, largely in the fringe, increased by forty percent. One of the main problems, especially with land owned by local authorities, has been the lack of funds. But the problem has also been compounded because public institutions and industries, such as the Central Electricity Generating Board, are not liable to planning restrictions.

Townscape: this is now scarred by deserted inner city areas and multi-carriageway roads. Coleman takes the example of the Isle of Dogs in London where businesses were encouraged to move out; it would be more attractive, said the planners, if congested industries left. But this served only to increase unemployment in the area. Rates for those remaining increased in turn. Planners here, as in other city areas, have over-reacted to urban overcrowding. They have concentrated on 'newness', on 'open space', and ignored the more subtle, but crucial, issues such as the age-mix of the community (there are often too many children in enclosed estates), on architectural design and street layout.

Coleman points out that research, notably by Oscar

Flats in Salford. *The loneliness of an urban nightmare. Priority was given to roads over green space, to "saving space" at a time when the amount of waste and derelict land increased dramatically.*

Newman in New York, has shown the importance of 'buffer space', really semi-public space, between home and the street. It could consist of a garden, or just a few steps. It is an area where strangers could be identified immediately when entering the community's, or the family's territory. Crime and vandalism has developed partly because, in modern estates, so much space, like alleyways, belongs to no-one but at the same time does not constitute a public highway. Open space is not treated in any functional way but merely as a concept, good for the sake of it.

In high-rise, large blocks of flats, people do not know each other, and inhabitants do not feel responsible for the areas. Council tenants are discouraged from making decisions; there is no scope for self-help. With tower blocks in one place, and a centralised shopping area in another, walking is positively dangerous; more trips are taken in the car. Attractive walkways are not even contemplated. And there are too few rooms per people in these estates.

Should not architecture, Coleman asks, be left to architects, rather than planners? Don't blame land speculators for high house prices, she says, since they can flourish only if there are shortages to be manipulated. Is it not true that inflationary land prices cause inflationary housing prices since when house prices are high, builders can afford to pay more for land?

Planners should be blamed for not giving top priority to housing people, and for concentrating too much on slum clearance and demolition. Delays in the planning machine, Coleman argues, have amounted to the equivalent of at least one year's housing starts, and this alone could make up the difference between shortage and sufficiency. Planning delays are estimated to cost the building industry £500 millions a year. Delays have led to premature evictions and demolition, especially by local authorities. Squatters have proved that many buildings from which original occupiers have been evicted are perfectly good dwellings.

The planning bureaucracy, concludes Coleman, is an "ever-burgeoning drain upon the economy".

Wildscape: this is the largely uninhabited area of moorland

and heath. Wild and natural it may seem, but it has valuable uses. The Ministry for Agriculture has admitted that it is possible to sustain a twelve-fold increase in the density of the sheep population on areas with a suitable mixture of heather and moorland grasses. Where poorer grasses, such as mat grass with no nutritional value, predominates the answer, according to Coleman, is afforestation. She points out that in our temperate climate, trees — conifers, at least — grow more quickly than in the Soviet Union and Canada where the virgin forests may be in danger of being exhausted.

Marginal fringe: this is mainly disused land, formerly good pasture but since invaded by bracken and moorland. Coleman has calculated that there are 2,000 square kilometres of bracken, an area equivalent to 1.4 percent of the land area of England and Wales. Moreover, bracken is a suspected carcinogen, causing stomach cancer in particular. (Japan, where bracken fiddleheads are considered a delicacy, has the highest rate of stomach cancer in the world. The poison can also be carried in water. Stomach cancer has been rising in Birmingham since the town began to use Welsh water.)

And she also makes the point that if the Ministry of Agriculture, in its estimates, includes marginal fringe which has been improved and upgraded into farmscape then it seems that even more high quality farmscape is being lost than the figures suggest. It implies that the overall loss in quality is greater than the overall loss in quantity.

The planners' gift to posterity is a massive stock of slum property and of wasted and derelict land. Delays and pressures in the planning procedures have also led to corruption. Clearer legislation and knowledge of people's freedoms are better, according to Coleman, than administrative wrangles between us as amateurs and the bureaucratic professionals. Why is there no legislation to preserve farmscape? Or to restrict tower blocks by law, or enforce obligations on public authorities, and make them liable for unnecessary evictions?

Planners are unaccountable. They should be named as the individuals responsible for specific proposals and decisions. That would allow the public to fix the blame where it belongs, enable us to tell who is responsible for derelict land. Maybe

there should be a kind of 'scoring' system based on clear
definitions of scapes, their functions and purposes.

Some of the more obvious requirements, says Coleman,
should be:

No demolition, and hence no eviction without default, and no
planning blight;

No high rise flats, and possibly no flats at all, since there are
already vast numbers accommodating people who would
prefer houses;

No estates of more than a stipulated size, to avoid faceless
urban monotony; fairly small gardens, to achieve urban
compactness and help restore viable urban densities;

No more council housing, since there is already more than
enough to house all the really penurious;

No industry that does not meet certain environmental
standards in relation to noise, pollution, smells, toxicity, etc;

No confused space. Ideally all space should be either private,
semi-private or public. Any semi-public space should be
controlled by a clearly recognisable community, such as a
small old people's home, with adequate provision for surveil-
lance.

"To emerge from under the yoke of planning would be
closely comparable to emerging from the feudal system",
Coleman says. "The feudal system prescribed the management
of the farmscape as minutely and often as arbitrarily as the
planning system has prescribed the management of settlement. It was the coming of more scientific understanding that
made it possible for each individual farmer to manage his own
farm without environmental disaster, and both production
and civilisation advanced as a result. Today we are developing
the scientific knowledge that will free us to manage our
townscape in more environmentally responsible ways. To do
so will be to advance civilisation, to fail to do so will be to
retreat into a modern kind of bureaucratic feudal overlord-
ship".

What is proposed is a national landuse planning strategy;

every area will be accorded its main role. It seems a sound and rational concept; land will be respected, and not wasted. But uneasiness then creeps in. A strategy is proposed, yet the need to counter bureaucratic, central, decision-making, the need for more flexibility and freedom is underlined. Who will decide on the strategy, especially if all the regions of Britain are supposed to dovetail in rational harmony?

Farmland owners want complete freedom from planning restrictions to do what they like. Given their collective vested interest, this means freedom to plough up land, build new sheds, cut down hedges. They enthusiastically welcomed the Thatcher administration's decision to repeal some 300 central government controls over the actions of local authorities. But this is a hopelessly short-sighted approach. What if other groups, the road lobby, for instance, or conservationists, demand the same freedom to use the land in a different way? New forestry plantations are often attacked by farmers as well as by conservationist groups but are not subject to planning laws now. The Ministry of Defence, one of the largest owners of farmland, can impose controls on land use without farmers having any say or right of appeal. Are farmers really happy to identify themselves with the attitude of Buckingham Palace which, when questioned about the decision to build tracks below Lochnagar on the highest mountain in the Balmoral estate to allow landrovers to assist those engaging in weekend shoots said: "as these tracks are being constructed on a privately owned estate, I do not see what business that can be of anyone other than the owners?" Are farmlandowners as content with the Conservative Government's decision that councils will not be bound by the advice of inspectors at public inquiries?

Public inquiries, in theory an instrument that could open up planning to popular investigation and accountability, are themselves fast becoming merely a tool to disguise the real prejudices and pre-determined decisions of Governments. Planning inquiries are extremely expensive. Protestors find it difficult to match the legal expertise — the QCs, solicitors, planning experts and their fees — which the Government, local authorities and nationalised industries readily have at their command. Farmers themselves complain bitterly at the

cost to them of fighting the National Coal Board's plans to dig
mines in the Vale of Belvoir. More significantly, the scope of
some public inquiries may be challenged. In a judgement on
the inquiry into the M40 and M42 motorways the law lords
under Lord Diplock ruled in 1980 that Government inspectors
have the right to prevent objectors challenging Whitehall
advisers on broad issues relating to Government policy. They
said the inspector, at the inquiry at Bromsgrove in 1973-74,
was right to prohibit the cross-examination of official witnesses
or the Government's method of forecasting traffic requirements. According to the Law Lords: "the decision to construct
a national network of motorways, necessarily in stages, is an
administrative decision and government policy, and it is not
open to question at local inquiries held to receive local
objections to a proposed scheme for a particular stretch of
motorway". Diplock went so far as to say that inspectors may
prevent challenges on issues even if these are not matters of
current Government policy. This has serious implications for
future inquiries. Expansion of nuclear power, for instance,
must be regarded as a foregone conclusion if questions about
the need for extra energy requirements and forecasts are to be
ruled out of court.

A national land use strategy, tied up by Whitehall, could
compound the problem. As Christopher Hall, former director
of the Council for the Protection of Rural England has said:
"In 1980 the environment secretary rejected a plan for the
Southern Water Authority to make a reservoir of the pretty
Broad Oak Valley in Kent, after an inquiry at which the
authority's forecasts of water demand were shot to ribbons. It
could well have been impossible, certainly far more difficult,
to do this if the scheme had formed part of a strategy
previously hallowed in Whitehall as the holy writ of national
policy".

Bureaucratic controls and decisions drawn up in offices
have, as we saw, deliberately kept the supply of houses below
the demand and hit the poorest hardest. Building societies
have imposed rigid criteria on both the type and location of
houses when deciding who should be granted loans. Planners
impose general standards which are inflexible and are not
relevant to the problems of individual areas. A road in one

place may be useful; in another, an expensive and even a dangerous nuisance. Council houses, or, at least, subsidised housing, may be needed, if we are basing our judgement on the existing system, for a large part of the population in one area, for the elderly in another, and for no-one in yet another.

The financial and environmental consequences of taking the easy way out in one specific sector have been described by William Tatton-Brown, former chief architect for the Department of Health and Social Security. He estimates that on the basis of plans and costings for two hospitals at Frimley and Bury St Edmunds drawn up by a small team of planners who knew the needs of the areas well, that the Government wasted at least £1,500 millions by constructing large, centralised, hospitals, during the massive building programmes of the 1960s and early 1970s.

Because decisions were taken by a group of experts on the basis of an ideal which looked good on paper, Greenwich hospital was equipped with air conditioning in spite of warnings by the nurses involved that they would want — and, indeed, have — to keep the doors open. At St Thomas' Hospital in London, small and single-room wards were installed. Doctors subsequently admitted that they were a mistake so far as the psychology of the patients is concerned.

"Planning", the euphemism for an abuse of freedom, affects all of us, and it affects us directly, our homes and our immediate environment. Yet instead of being properly founded on the criteria of human need and human values, it has succeeded only in ignoring our basic interests. It has been prey to the market and the market, in turn, is subject to capitalist forces, and priorities. There are no social controls, or guidelines for developers. The right to a home is a fundamental one. People should be allowed immediately to build their own house with materials provided out of the profits of those organisations, institutions and companies that have made so many millions of pounds depriving the rest of the community of land, space and homes. It is highly significant that in any boom, entrepreneurs have swiftly moved in to speculate in property, and aimed especially at office blocks which provide no-one with a home or a garden or recreation facilities, but only the lure of quick and high rents from an unproductive sector of the economy.

It is a story of waste and greed. The more land that is made idle or abused for short-term business operations, the greater the number of homeless and the amount of farmland under threat. Meanwhile, far away investors are given an excuse to cut down the Amazon jungle. Nearer home, they also tear up the countryside, encouraged by the spurious argument that unless more land is ploughed, people will be short of food. That, of course, is nonsense. There is plenty of land. It is all a matter of how the land is used.

This viscious circle must be broken, but it needs an attack on all fronts — questioning our attitude towards health, diet, land use, trade, transport, planning — if we are going to be successful. Above all, it must be up to individual communities, and then to discussion between communities, to decide. Token gestures of participation must be rebuffed; a change in the power structure is essential.

A few precedents have already demonstrated that initiatives can be taken even within existing legislative framework. In the 1940s, a housing association was set up in Blakeney, Norfolk, to ensure that houses should be reserved for local people. More recently, the Gwynedd Housing Society and the Lake District Special Planning Board have limited the use of new housing to local inhabitants. A section (no. 52) of the Town and Country Planning Act established the needs of local people and local employment as criteria. But these examples are rare exceptions to the rule. A dramatic change is needed if the conflict, growing day by day, is to be resolved.

With the election of the Conservative administration in 1979 the political climate changed. The Environment Secretary, Michael Heseltine, made no secret of his impatience with traditional bureaucratic procedures. But the new priority, while questioning past planning legislation and its ineffectiveness, mainly promoted the interests of developers. The crux of the Government's new planning philosophy was contained in the Department of Environment's circular 22/80.

When he introduced the circular, Mr Tom King, the Local Government Minister, said: "Hitherto, there has been too much emphasis on restraint and restriction. From now on we intend to ensure that positive attitudes prevail". Local authorities were asked to pick out for priority handling "those applications which in their judgement will contribute most to national and local economic activity".

Authorities, faced with growing economic pressures, were poorly placed to resist the march of commercial developers, particularly if developers offered councils large capital payments or leasing arrangements, an increasingly popular practice. Those who resisted these pressures were quickly put in their place. The Department of Environment overturned 5,000 decisions made by local authorities against developers in 1981, compared to 3,000 in 1979. Much greater 'tolerance' was given to residential and industrial development and applications were processed much more quickly. By 1981, Heseltine used his direct power to increase housing allocations beyond those made in local plans in Berkshire, Buckinghamshire, Gloucestershire, Devon, Wiltshire and Hampshire. He instructed his officials to make greater use of Special Development Orders to enable, as the Council for the Protection of Rural England (CPRE) put it: "broad classes of new housing and industrial development to proceed effectively by the Secretary of State's fiat, outside the development control system".[1]

Green field sites were threatened. The CPRE warned that the Government was encouraging the piecemeal erosion of the countryside under the guise of efficiency and quick planning decisions. The planning climate had been fundamentally altered to favour developers at the expense of widely accepted policies of restraint. There is a very real prospect of small market towns and villages merging with each other, with both housing and industrial developers destroying green field sites as companies and individuals moved from inner city areas to more attractive and increasingly accessible country areas.

And the Government's attempts to attract investment to inner cities seduced embarrassing projects. With the London docklands still offering the promise of employment and new communities, one of the first initiatives was a plan in the Limehouse Basin for a complex of restaurants and cafés, with possibly a marina attached.

Narrow Boats lying desolate in a canal at Norton Caves, nr. Cannock, Staffs. *They used to take coal, but they are another reminder that inland waterways used to play a vital part in Britain's transport system. They could be cleared at relatively little cost. (Photo: Press Assoc.)*

9

CONFLICT

"The heart of our urbanised England is still in the country" – *G.D.H. Cole on the life of William Cobbett.*

"The countryside cannot be treated as just a huge food factory" – *the Ramblers' Association.*

"You may be asked to vote on such issues as land nationalisation, leisure and sporting access, conservation and preservation of the countryside. Remember that the farmer's land is his factory where he produces food" – *Colonel Mackain-Brenner, Wiltshire secretary of the Country Landowners' Association to a group of schoolchildren.*

In the towns, the conflict between speculators and planners, developers and public authorities, between bureaucrats and the individual leaves in its wake dereliction, poverty and an ugly environment. In the country there is another growing, starker, conflict and conflicting pressures: the search for recreation, space, clean air, property, increased food production, coal and other minerals, cheap foods, dumps for nuclear and other industrial waste, more arteries for traffic, preservation of wildlife and increasingly rare flora and fauna.

The hitherto complacent approach towards the gradual polarisation between town and country is in danger of turning into an explosive controversy between what increasingly is perceived to be a small group of privileged landowners, free from planning restrictions yet benefitting from tens of millions of pounds of public money in subsidies and tax concessions and the vast majority of the population living in towns and cities.

The town/country apartheid is unique to Britain, and the countryside is under threat as never before from interests whose one common denominator is the search for money and the pursuit of profit.

Many farmers and landowners do love the countryside and appreciate its beauty. Members of the Landowners' Association are busy planting trees while their neighbours — farm managers, perhaps, for large companies or financial institutions — dig up hedges and ancient copses. Small farmers sweat to earn a meagre living and having nothing to spare. But for all of them, the pressure to sink under the tide of competitive capitalism or even run, lemming-like, over the cliff and say "to hell with the land" is mounting day by day. And all the time more and more people in large towns want to free themselves from the harsh materialism and jungle priorities of urban life and from the intolerance of municipal bureaucracy. The pressure intensifies, as they want to get away from the squalor, noise, and claustrophobia of the towns. "While you can buy almost every kind of luxury and convenience," as one commentator put it, "the time comes nearer every day when it will be impossible to buy the ultimate luxuries of all, privacy and silence".[1]

The central conflict is between the farm landowner, property speculator, politician or bureaucrat who regard the land as a source of revenue, something to be exploited, and the others who rightly recognise it as a precious resource. Behind this conflict lies a more fundamental danger, the prevailing attitude, the arrogance, which suggests we are superior to Nature, that we can do what we like to the land. We do not yet realise how easy it is to destroy the ecological balance which takes generations to restore. And even though the evidence is building up to a startling degree, we do not take it seriously because it does not touch us personally. We live in an increasingly impersonal world, divorced from our neighbours, let alone our roots in Nature and the land.

In the conflict over the land, the conservationists are in the vanguard, as they should be. The Countryside Commission, a body more successful in its analysis than in its influence, has described succinctly one cause of this conflict, namely the loss of wildlife and plant habitats, itself one of the clearest manifestations of how the land is being abused. "The adoption of new farming systems and technology", it says, "over the past 25 years has been encouraged by the Ministry of Agriculture, Fisheries, and Food through a series of measures designed to keep food cheap by promoting greater agricultural production at lower costs: draining of wetlands, enlargement of fields by hedge removal, the ploughing up of permanent pasture and heathland, and investment in new farm buildings have all been encouraged by government grants and advice . . . a habitat that has taken a thousand years to evolve can be bulldozed, burnt and ploughed for cropping, in a day".

The Ministry spends about £500 millions annually to farmers in grants and subsidies. Some £150 millions of this is earmarked specifically to the destruction of valuable habitats. On top of this, the EEC spends about £650 millions a year on supporting British agriculture through grants and price guarantees. British taxpayers and consumers subsidise each farmer to the tune of thousands of pounds each year and farmers now receive more than half their income from central government. A mere £10 millions is spent on conservation. Yet shortage of food, the argument constantly employed by both farmers and Whitehall, is not the problem: indeed, because of high EEC prices, farmers are switching to products, chiefly cereals, of which there is a massive and increasing surplus in the EEC.

The lack of discrimination in the way the Ministry hands out grants was dramatically illustrated when a visiting delegation from West Germany asked Whitehall officials whether they would have given a grant to Paul Getty if he had applied for a grant to increase the profitability of his land. The officials replied: yes, they would.[2] As Marion Shoard has said: "farmers' means are not examined before they receive a grant — as are those of citizens seeking legal aid or rent rebates, for instance".

And while food may appear relatively cheap in the shops, the cost to the countryside, to wildlife, and to farmers faced with rising fuel bills, of the methods whereby food is produced is not worth the price we are paying.

The Ministry of Agriculture has a legal obligation to "have regard to the natural beauty of the countryside". But it makes no secret that it sees its role only as one of promoting the interests of food producers. "Our policy is one of keeping our heads down", said one senior Ministry official during the debates on the Wildlife and Countryside Bill in Parliament in 1981. Instead of increasing the role of its Agricultural Development and Advisory Service (ADAS) to encourage farmers — who are themselves beginning to question current agricultural methods — the Government in 1980 reduced the number of officials engaged in this work. Agrochemical companies could have no better spokesman that Keith Dexter, the ADAS director-general. And in order to reduce public expenditure, the Government announced that henceforth farmers can go ahead with development plans eligible for grants without first having to consult the Ministry. They are asked to consult authorities, such as the Nature Conservancy Council (NCC), before developing land in officially-protected areas. In practice, of course, once a landowner has gone ahead, even if he eventually does not get approval for the grant, the damage will have been done.

It is useful to establish a kind of catalogue, by no means exhaustive, of the unprecedented damage already done to the countryside over the past years. It is a picture the casual visitor to the country may not easily see or appreciate.

Lowland areas in particular, including chalk downland in the south, grassland, wetlands and heaths have been ploughed up, cultivated and drained. Moorland has been attacked, partly by plantations of conifers. Semi-natural woodlands are fast disappearing. Here are some examples of the loss of natural and traditional habitats:

140,000 miles, nearly a quarter of the total, of hedgerows have been dug up since 1945. Over the past twenty-five years, about 80 percent of the hedgerows in the arable, heath and downland areas of the south and east have vanished. Norfolk, a large county and one of the very few where agriculture remains the dominant sector of the economy, has lost nearly half of its hedges — 8,000 miles in all.

A half of all Britain's deciduous, that is to say, broad-leaved, woodland has been lost since the war. In Scotland, traditional woodland now accounts for only about one percent of the land area. Woodlands, and in the case of Scotland, large areas of heathland as well, have been taken over by commercial forestry plantations of conifers. Over the past thirty years as much traditional woodland has been lost as had been destroyed or felled over the previous four hundred years. For instance, in 1977, Gladdle Wood, near Dover, 16 acres of ash and maple and the home of rare orchids, were flattened. (See fig 15, p95)

Some 15 percent of Britain's old meadows, a source of such increasingly rare flora as the snakeshead fritillary, and of which many have remained fundamentally the same since the Middle Ages, have been ploughed up or drenched with chemicals, their character totally changed, over the past two years. In September 1980, about 30 acres of Wendlebury Mead meadow in Oxfordshire, a grade 1 Site of Special Scientific Interest (SSSI), were ploughed up. A similar fate befell Long Meadow at Ashton Keynes, Wiltshire, another SSSI, five years earlier. At the time of writing, Frays Meadow, part of the Colne Valley in west London, an SSSI and a Nature Reserve and the last remaining ancient flood meadow owned by the Greater London Council is under threat from mineral extraction to provide £4 millions to spend on what the GLC described as "environmental treats", such as life-size plastic dinosaurs in London parks. The meadow has been a home of kingfishers, herons and many species of butterfly.

75 percent of Suffolk heathland has been lost since 1920, 42 percent of the Dorset heaths have vanished since 1960. Hundreds of acres of heathland have been ploughed up or developed in Southern Scotland, East Anglia, Hampshire and Surrey over the past few years.

Over a third of Dorset's chalk downlands have been ploughed up since 1967. The amount of downland in Wiltshire has been reduced by about half since the war, those acres not attacked by the plough being mainly military training areas. Over the past fifteen years, about 20 percent of chalk grassland on the Isle of Wight in Sussex and Hampshire have been ploughed up. The case of Graffham Down in Sussex has been documented in detail by Shoard[3]. Bought in the 1970s by two Dutchmen, the van de Vegte brothers, 83 acres have been ploughed up with the help of a 20 percent grant (amounting to about £4,000) from the Government to grow barley, of which there is a surplus in the EEC.

Of the 120 important grassland sites listed in the Nature Conservation Review, fifty have suffered serious loss of scientific interest in the past ten years. Only about a third of the rough grassland and heaths that existed in Devon at the beginning of the century, remain.

Mosslands, wetlands and bogs — particularly valuable sanctuaries for wildlife — are being drained and reclaimed faster than ever by farmers who often fear that a protection order may be placed on these parts of their land. About 150,000 acres of wetlands are drained each year at a cost of over £20 millions, the bulk paid by the taxpayer through Ministry of Agriculture or EEC grants. The land is then used for pasture for milk production while pasture and grassland elsewhere is ploughed for cereals, though there is a surplus of both products in the EEC, falling consumption of milk, and increasing exports of grain.

The Wash and its saltmarshes are of enormous international importance, with about ¼ million birds wintering or resting there. But it is getting smaller; almost half of the original saltmarsh there has been reclaimed. The Wash is a source of nutrients, mainly protein. With the support of the Ministry of Agriculture, farmers want to drain about 200 acres at Gedney Drove End, on the south-west shore of the Wash, to grow wheat, barley and sugarbeet for which there is no domestic market. In the event, common sense has temporarily prevailed, and Michael Heseltine, the Environment Secretary, was advised to reject the proposal after an inquiry.

The Somerset Levels, one of Britain's largest wetland areas, rich in wildlife, flora and fauna, are being threatened by landowners, encouraged by Whitehall and the prospect of EEC grants. Farmers want to drain them to introduce intensive arable cultivation and livestock farming. The land is extremely fertile, 20-feet deep in peat in some places and worth more than £2,000 an acre. Farmers insist that it must be used to its full agricultural potential. It has been farmed since the thirteenth century by small, extensive, livestock holdings. But unlike these farming methods, controlled by human labour, modern agriculture, dominated by chemicals and machines, is not adaptable to the particular characteristics of different soils or sensitive habitats.

Since the war, ninety percent of 169 square miles of the Levels have been drained. In 1979, the Ministry of Agriculture offered grants for the draining of a further 40 acres of West Sedgemoor. Then, in 1981, a landowner, Stanley Durston, cleared trees from a peat bog, part of Street Heath Site of Special Scientific Interest, near Glastonbury, to try and force Somerset county council to give planning permission to dig up the peat. Permission was granted since the site's conservation value, the trees, had been destroyed.

In 1980, Lord Thurso, the Liberal energy spokesman, applied to the Ministry and the Forestry Commission for grants to destroy half of the extremely valuable peat bog of Blar nam Faoileas, described by the Nature Conservancy Council as a rare and almost pristine eco-system. Landowners have drained stretches of the Halvergate Marshes in Norfolk, ignoring new legislation whereby they should first have notified conservation agencies.

About 20 percent of Exmoor has been ploughed up since it was designated a National Park in 1954. The Government now says that the problem can be contained by special management agreements between conservation agencies and the land-owners. About the same proportion of the North Yorkshire Moors have been reclaimed, ploughed, or fenced off since they were designated a National Park in 1951. The story is similar in Snowdonia and the Brecon Beacons. According to the best estimates[4] Britain's moorlands are being enclosed or reclaimed at an average rate of 12,000 acres a year.

Many acres of moorland now being ploughed up have not seen the plough for 1000 years. As a recent study pointed out[5] they are the nearest we have to 'natural' landscape. And judging by past experience, many other acres of what is, after all, marginal land, will revert to moorland.

Habitats are not worthy of protection for their own sake; many different species of wildlife depend on them. Birds of prey — the merlin, or peregrine falcon — need open moorland and are threatened by afforestation. A typical fir tree supports sixteen species of insect life; an oak tree, 284.

Greenshanks, golden plovers, dunlins, the Dartford warbler and the nightjar are all threatened with extinction. The last two birds, like the sand lizard, smooth snake and natterjack toad, rely on heaths and are now endangered species. The snakes-head fritillary and the marbled white butterfly depend on traditional meadows. Wetlands and bogs are the only habitat for marsh harriers, kingfishers, marsh saxifrage, the fen orchid, the lady orchid, the bogbean, the bearberry and marsh fern. The cowslip, the common toad and common frog, are no longer common. Long lists of species — the chequered skipper butterfly, the mole cricket, the fen raft spider, greater yellow-rattle, field cow-wheat, least lettuce among them — were introduced by worried peers in the debate in the Lords on the Wildlife and Countryside Bill in 1980 and 1981.

Nine of the twenty-six species of bumble bee have seriously declined in southern England. Does it matter? Yes; it is a warning of things to come, of the way we are going down the road towards uniformity. Uniformity dulls the spirit; it is impoverishing. Most of all it encourages diseases. Some habitats can never be re-created. Their destruction snaps our unbroken link with the past. Nature may be resilient, but it can do nothing against chemicals and the bulldozer. How does all this marry with landowners' claims that they are stewards of the land, responsible for future generations?

It is quite obvious that existing arrangements designed to protect valuable habitats are not working and that, with present trends and a voluntary system of agreements with landowners not effective, only publicly-owned and administered nature reserves along the lines perhaps of the system in the US and Canada, will protect sites effectively. 10 percent of all Sites of Special Scientific Interest are lost each year; a third of those in Dorset were destroyed in 1980 alone. The Government acknowledged in 1980 during the debates on the Wildlife and Countryside Bill that of the 735 'special interest' sites said by the Nature Conservancy Council (NCC) to be the most important, over 50 are seriously damaged and hundreds more are threatened. Graffham Down was officially designated an area of outstanding natural beauty; Wendlebury Mead meadow was a Grade 1 SSSI. A 450-acre grade 1 SSSI, Horton Common in Dorset, is being ploughed up as I write.

No notification or planning permission has been required for agricultural or forestry development in SSSIs. Landowners complain that they do not know whether or not part of their land is designated as a valuable habitat. When they do know, they can go ahead regardless, as we have seen in the Somerset Levels. Farmers do not even have to consult the NCC. A farmer in Mellis, Suffolk, early in February 1981, destroyed a 1000 year-old copse, a haven for nightingales and woodpeckers, just a day after the local newspaper announced a decision by the local council to introduce a conservation order on his woodland and hedgerow.

There has been no effective preservation machinery controlling the status either of SSSIs, Areas of Outstanding Natural Beauty or National Parks (which in theory account for over 20 percent of England and Wales). There are no effective consultation procedures between their owners, those organisations which designated the areas in the first place, or local authorities. The system is in a shambles. The NCC frequently fails to notify local councils or landowners where it has designated a site (if a site is so designated, the authority must consult the NCC before granting planning permission for new developments). If it does seek approval from the landowner and does not get it, the NCC has tended to give in, relying, sometimes, on 'gentlemen's agreements'.

As Catherine Caufield, in one of a series of articles early in 1981 in the *New Scientist*, pointed out, the NCC has not designated for conservation Llanbrynmair Moor, in Wales, though the Royal Society for the Protection of Birds describes it as an important breeding ground for merlins, golden plovers and greenshanks. Parts of the Moor are being bought up by private commercial foresters who are claiming grants from the Forestry Commission for conifer plantations. The NCC delayed taking a decision on the Berwyn Mountains, though it admitted it should be classified as a SSSI. An untidy compromise after months of bitter argument between farmers, foresters and conservationists finally led to some 69,000 acres being designated for protection, for merlins and hen harriers in particular. It was a proposal which satisfied nobody.

The NCC owns only thirty-five of the 167 Nature Reserves. Its financial resources — £9 millions in 1980 — are totally inadequate for its task. The potential financial demands on the Council were dramatically illustrated when it bought the Ripple Estuary for £1.7 millions in 1979 to prevent it from being drained and used for arable crops. It had earlier been bought for £1.2 millions by a Dutchman who planned to reclaim the estuary with the help of Ministry of Agriculture grants. As Gerald Wibberley, former Ministry official, NCC member, and head of the department of agricultural economics at Wye College put it[6] "we paid the full development value for the most intensive possible use of that land. I voted for the purchase in the hope that it would bring home to the nation the high opportunity costs involved in safeguarding wildlife in areas with the potential for intensive agriculture. What a vast sum of money we had to pay just to preserve an area of marsh". And yet the only sanction the NCC has is the threat to buy up property or land.

Much of the blame for the lack of effectiveness and coherence surrounding the whole issue of preservation must be laid at the door of the two and official umbrella organisations responsible for the countryside in Britain. And one of the problems, as Ann and Malcolm MacEwen[7] describe is the completely artificial separation of the role of the Countryside Commission on the one hand, and of the NCC on the other. The former concentrates on recreation, access to the country, preservation of the beauty of landscape, on scenery. The NCC concentrates on the preservation of 'nature'. The two issues are, of course, inseparable; a fact recognised in other countries. Wildlife, flora and fauna cannot be separated from their habitats which in turn make up the landscape.

But it has meant that the civil servants on the Countryside Commission have dwelt, not very effectively, on the pressures on the country and on designating special areas, while the NCC has concentrated on more specific scientific interests, distancing itself from the political arena. The Commission has taken on the task of facing the problem of access and amenities; the NCC has not.

These different approaches have compounded the existing weaknesses of the two agencies. This can be well illustrated by two sites in the Brecon Beacons. Llangorse Lake and the afforestation of part of the Cnewr estate. The lake is a Grade 1 SSSI, yet the NCC has failed to carry out a scientific survey into the causes of its gradual destruction. The Countryside Commission, in conjunction with the Sports Council and Brecon County Council did conduct a survey recently, but with recreational facilities its main concern. It showed that the lake was coming under tremendous pressure from competing consumers — including canoers, sailors, water skiers. Nothing happened. The Secretary of State for Wales insisted that the management of the lake should be left to voluntary arrangements. Meanwhile, the local lord of the manor, who owns the lake bed, prevented the creation of a nature reserve. Commercial companies promoting adventure holidays and so on successfully resisted attempts by the National Park Committee to enforce planning decisions despite several enforcement notices backed up by Whitehall.

On the Cnewr estate, where a compromise on afforestation was finally accepted by the National Park Committee, the MacEwens made an estimate of the amount of money owners could get in grants from the Forestry Commission and the Ministry of Agriculture by investing in new plantations and sheep production on 775 acres. Total state aid over a period of ten years on such as estate owned, for instance by the Economic Forestry Group, a private organisation, would total £286,500. But the very most the National Park Committee could have offered as an inducement to preserve the existing landscape would be just £500. "The conservation of natural resources (must) be built into the entire apparatus of subsidies, grants, tax incentives, farm development plans and dedication agreements for agriculture and forestry", the MacEwens conclude. Secondly, they urge all agencies and voluntary bodies to unite and increase their influence with those who matter — the landowners. As we have seen, there is now no effective or coordinated procedure governing consultations between local authorities, pressure groups, the Government, farmers, let alone an effective means of control and of articulate decision-making on the basis of a debate about priorities.

Farmers dig up trees and hedges for a simple and fundamental reason: to increase production and therefore profits or income. Small fields — which are ideally suited to rotational grazing, for instance — are regarded as a nuisance, especially by arable farmers. The department of Land Economy at

Cambridge University[8] has described in great detail how a landowner can save, in terms of wages, machinery costs, and in tractor and combine time, by enlarging his field. (Machinery costs and time would be reduced mainly by cutting down on the number of turns.)

At first sight, the temptation seems obvious. A farmer with 500 acres who doubles his average field size to 50 acres stands to save nearly £7,000 a year for the first six years, according to the Cambridge study. This includes the loss of one farm-worker, 'saving' £3,000 a year. But this sort of criterion is not only short-sighted; it reflects a superficial, mean approach to agriculture and exploitation of the land. It positively encourages rural depopulation and unemployment, and the move to intensive, specialised farms. It leads to that ugly and materialistic trend towards uniformity of the landscape and farming practices.

It also illustrates the fundamental problem of how to discourage farmers from looking after their land in the interests of good husbandry and of rural communities, and indeed, of the common good. The Conservative Government opposed the introduction of any compulsory powers or rules restricting farmers' freedom of action, during the debates on the Wildlife and Countryside Bill in 1980 and 1981. Its attitude was best summed up by Jerry Wiggin, then Parliamentary Secretary at the Ministry of Agriculture, who told the Commons: "there is no such thing as the natural beauty of the countryside... it is totally untrue that farmers and landowners are destroying the countryside".

Designated areas should not benefit from statutory protection, either from central or local government, according to Ministers. They agreed that development of SSSIs and nature reserves should at least be postponed pending consultations during a period of between 3 and 12 months with the NCC and local authorities. Moors could also be protected for a period of 12 months but only if a specific ministerial order has been made: something the Government made it clear would rarely, if ever, be done. But local authorities or the NCC will only be able to save protected areas, under the terms of the Act, if they had the money to buy them.

Under the Wildlife and Countryside Act which became law

at the end of 1981, when the NCC establishes an SSSI, it has a duty to inform the landowner what action by him could be damaging. If the owner wants to go ahead and develop the land he has to give the NCC three months' notice. If the NCC objects to the plan, it would endeavour to reach a management agreement with the landowner, possibly including financial compensation. If the farmer ignores the NCC's views, he can be fined up to a maximum of £500.

Areas of Outstanding Natural Beauty are not covered by the Act. The Act is based on the hope, though practice in the past scarcely warrants it, that the country can be looked after by voluntary cooperation between landowners and the rest of us.

There is no provision in the Act for compulsory fallback powers. The Government defeated a proposal that the Ministry of Agriculture could pay grants for such generally important objectives as reviving the rural economy or conserving the countryside. So Ministry grants will be paid solely for agricultural development and the Ministry will be required to promote conservation only if this is consistent with the farmer's interests and with the promotion of 'agricultural business'. The Government also used the Parliamentary whipping system to reject an amendment whereby if a farmer is refused a subsidy on conservation grounds, compensation would be permissive and discretionary, not mandatory.

But the most significant and alarming feature of the Wildlife and Countryside Act is that it has introduced an entirely new principle into the whole question of agricultural landownership. It says that if a grant is withheld, or if a landowner agrees not to go ahead with a drainage scheme or felling woodland, for instance, then he will be entitled to automatic financial compensation.

Landowners, therefore, have the right to compensation, paid for by the taxpayer, simply for placing the public interest — the public's rights to the land and its conservation — before their own private interest. Moreover, conservation agencies, which rely either on voluntary contributions or Government funds, will have to pay the compensation fees. One of the first agreements, involving the Countryside Commission and the Norfolk Broads Authority, led to nineteen farmers being paid £65,000 a year not to drain their land on the Halvergate Marshes. The cost of compensation could run into tens of millions of pounds, perhaps up to £100 million a year if all nature reserves are going to be protected. Yet the budgers of the NCC and the ten National Park Authorities together amount to less than £20 millions.

Otters, according to the Act, would remain protected species, yet there would be no ban on otter hunting. Coniferous trees such as Sitka spruce will be able to be planted anywhere in the wild. Landowners will have a licence to kill birds of all species if they were damaging, or thought to be damaging, crops. Hector Monro, combining his role then as Minister for Sport with Minister responsible for the Act in the Commons, is himself a keen wildfowler.

The most persistent theme running through the Act is that voluntary codes of conduct and gentlemen's agreements are the answer. This flies in the face of all the evidence, and the widely-held view of scientists and others who have observed closely what is happening, namely that valuable habitats — for wildlife, flora and fauna or even the human eye — can be protected only if they are shielded by and can rely on some form of public planning control. There are, of course, landowners sensitive to the needs of wildlife and even of the public good, but we can no longer afford to ignore the need for a radical change in the prevailing attitudes as much as in the system, and face the fact that most decisions now are dominated by the search for more money, with the land treated only as a potential pot of gold or a last vestige of security.

Most farmers — and certainly their interest groups, the National Farmers' Union and the Country Landowners' Association — argue that the landscape is man-made and has been so for hundreds of years. It has constantly changed, they say, as a result of the development of agricultural methods and husbandry, of climate even, and of economic conditions. Fenland in East Anglia was once a large oak forest; Southern downland was ploughed by the Anglo-Saxons. There were fewer hedges before the Enclosures than there are today. Farms and farmers are not there just to preserve a picturesque way of life so that urban cousins can come and gape at them out of the car at weekends. So the argument goes.

"It is mad," says a Yorkshire farmer, "that urban people

have become the planning authority for the countryside and that planning decisions originate in town and city councils". The fact remains that in many rural areas, farming interests are well represented, even dominate, and frequently combine to form the largest single group on local councils. Private landowners successfully block plans that would provide jobs or alternative housing for this would threaten their control over the area and, in particular, loosen their grip over the labour market.

Whether they acknowledge it or not, large landowners sometimes form an alliance with conservationist groups to preserve the status quo in the countryside. They are less than honest when they accuse conservationists of wanting merely to preserve the wilderness without working in it. Sometimes they are joined by the local population which is equally hostile to any attempt by the outside world to share their space. Local residents, for instance, strongly supported farmers against plans for a Country Park on the site of the Battle of Hastings, insisting that there was no public interest in the scheme and that the Department of Environment was simply creating a demand where none previously existed. (This alliance between local inhabitants and farming interests is not consistent, of course: rural residents, especially those whom farmers refer to as 'nouveaux countrymen' from the towns, object to farm development, especially those pig and poultry farms whose main nuisance is the smell.)

Farmers are attempting to ban rock-climbing on the Roaches and the Five Clouds, a gritstone outcrop in the Peak District National Park. The Park Authority could not afford to buy the surrounding land; farmers brought in sheep to graze there and claimed compensation for damage as the climbers, they say, leave debris and cans which then harm the stock. It is an example, sheep worrying by dogs is a more common one, where lack of consideration by some give landowners excuses to prevent access to the countryside to all. So, as farmers erect 'no entry' signs on paths through woodlands, on gates to fields, they argue that preservation and access is a contradiction in terms. What many of them really mean is that they want the freedom to treat their land as private property, much as one treats a home, garden, or car.

It should be recorded that notices stating that 'trespassers will be prosecuted' are very misleading. Prosecution for trespass does not exist in law. Charges can be laid only for malicious damage. Owners may sue for damage, but courts in the main still have little sympathy for landowners who sue in cases of harmless initiatives by people entering his property without permission.

Farmers are increasingly hostile to pressure for greater public access to the country and to any move which suggests that restrictions should be imposed on modern farming and agricultural developments. The Countryside Commission wants to ban all vehicles from the Ridgeway, the 86-mile ancient path stretching from the Thames valley to Avebury, Wiltshire, and beyond. The public wants access; local farmers complain that cars and tourists damage their crops and make it difficult and expensive for them to carry on with their agricultural activities. They say that only farm traffic should be allowed access to the Ridgeway. The Countryside Commission, meanwhile, says that tractors above all must be blamed for the deep ruts damaging the path. Farmers say that if tractors are banned three acres per mile would be lost to food production.

The whole concept of National Parks received a severe drubbing in 1973 when a plan to designate the Cambrian Mountains in mid-Wales the eleventh Park was abandoned in the wake of opposition from local farmers. Why, they asked, designate an area which is a natural wilderness? By the same token, farmers were the first to welcome the decision of the Countryside Commission not to designate the Norfolk Broads a National Park and, instead, try and encourage a consortium of local authorities to monitor the problems of pollution, overcrowding and ecological damage. (The plan by farmers to drain over 5,000 acres of the Halvergate marshes on the Broads came to light only because the Broads Authority persuaded the Ministry of Agriculture to give it prior notification of drainage schemes. The initiative was designed to allow landowners to switch from grazing to more profitable arable farming. It would have had a serious impact on the last remaining stretch of the Broadlands' open marsh grazing landscape.)

The debate is full of contradictions. Farmers say that designations of protected areas threatens their freedom and, in particular, attracts the public to these areas. Yet, as we have seen, designation has had little effect on agricultural policy constraints, on the granting of Government subsidies, on forestry, conservation, planning or the construction of roads. A third of Anglesey, for instance, consists of Areas of Outstanding Natural Beauty, yet the local authorities allow heavy industrial investment there, while at the same time ban caravan sites. We can give many other examples of anomalies.

The M25 motorway will run straight through the Kent Downs AoNB and while landowners and the Council for the Protection of Rural England are opposed to the road scheme through the Derwent valley in Kent and Surrey because it will eat up farmland, the local villagers are in favour since it will save their homes from the effects of commercial and tourist traffic. National Parks are not safe from mining (though the Friends of the Earth did succeed in preventing Rio Tinto-Zinc from mining copper in Snowdonia — an exception). And, significantly, a dispute in 1979 over whether the North Pennines should be made an AoNB centre on concern that rural depopulation was already severe — the implication being that a designation would hinder future investment and employment possibilities.

Farmers bitterly opposed the proposal for a 260-mile footpath — a 'Cambrian Way' from Cardiff to Conway — on the grounds that this would lead to further erosion in Snowdonia and Brecon. Yet at the same time, there are growing economic pressures on hill farmers in the region to attract and accommodate tourists — including caravan-letting on farmland — to supplement their incomes.

But there is a real possibility that the only solution will be a decision to earmark a limited number of specific areas 'honey pots' for access to the general public. "There will usually be a need", the Royal Institution of Chartered Surveyors, says, "to confine the public to specific, well-chosen and well-serviced areas from which large parts of the AoNB could be seen although not entered upon except by public footpaths, bridleways and other public rights of way".

But this, too, could threaten and spoil those areas. The

Mountain Centre in the Brecon Beacons National Park, designed to give information to walkers and climbers, became a target for motorists. It received as many visitors on the first day it was opened as were expected in a full year. Parts of Snowdonia, which has to cope with half a million visitors annually, is crumbling; Hadrian's Wall is also crumbling; local authorities urge visitors not to climb the Malvern Hills but visit, instead, the surrounding countryside which, they say, is just as attractive. More than 16 square miles of the Peak District National Park have been seriously eroded and a further 200 square miles are threatened. It is being attacked by air pollution; lead, zinc, copper and sulphur dioxide from the cities and acid rain; by fires, sheep grazing, the general climate, and by people. Feet have widened the Pennine Way by forty percent in 10 years.

Stonehenge is visited by about 800,000 visitors a year. The stones have been scratched. The chalk long men and white horses in Southern England have also been vandalised. Scientists complain that Stonehenge has been roped off. Meanwhile, Woodhenge, about ten miles away, attracts hardly any visitors.

The Suffolk Breckland, one of two remaining habitats (the other is a secret location in the Chilterns) of the Military Orchid is protected like a military camp. The nature reserve is surrounded by a six foot-high barbed wire fence and open to the public for just five hours on one Sunday each year when the plant is in flower. It has been threatened by the cooling of Britain's climate, but also by afforestation and the ploughing up of chalk downland. The Orchid's tiny seeds need a south facing chalk bank with a small, creeping fungus called mycorrhiza. The old quarry in Suffolk used to be the site for 2,000 or more Military Orchids (so-called by Linnaeus because the flower resembles a medieval soldier's helmet) but many died because conifers planted by the Forestry Commission blotted out the sun during crucial afternoon periods.

Water is another source of conflict. Farmers want to reclaim land from the sea, drain marshes. Conservationists are intent on preserving wetlands while water authorities want to take more farmland to store more water. The Central Water

Kielder Dam: This, the largest man-made lake in Europe opened in 1982. There is now a large excess of water storage in the North, and water that is now wasted in homes and industry could be saved at a fraction of the cost of this reservoir. (Photo: Northumbria Air Fotos)

Planning Unit estimates that consumption of public water supplies will have increased by 17 percent between 1974 and 1981, and will increase again by 65 percent by the year 2001. Water authorities want to build a new reservoir at Okehampton in Devon. The Severn-Trent Water Authority plans to dig a new reservoir at Carsington with the loss of 1,000 acres. There are many other examples, as many as there are examples of water leaking from reservoirs badly in need of repair or poorly designed and constructed.

Reservoirs provide amenities such as boating and fishing. But in an unusually enlightened initiative on the waterfront, the Ministry of Agriculture (under a Labour administration) decided in July 1978 against approving a £300,000 plan to drain the Amberley Wild Brooks in Sussex, a 900-acre area which is waterlogged in winter months. The decision followed a public inquiry in which 13 conservation groups gave evidence. The Brooks support thirty-five types of grasses, some of them rare, and in winter Bewick's swans, Canada geese, shoveller, widgeon and teal. Conservationists prematurely welcomed the decision as a turning-point. "Rubbish," replied the National Farmers' Union. "All the nature conservancy bodies are to be stopped from coming on our land." Because of the Brooks, it added, farmland was being flooded and farmers would have to be compensated as a result of land being conserved.

More consistent with our theme and our warnings, British Nuclear Fuels Ltd. planned to take water out of Wastwater, Britain's deepest lake. The North-west Water Authority said that the company should take water from Lake Ennerdale instead, a decision that would lead to a significant loss of farmland in the Cumbria National Park. BNFL planned to draw up 16 million gallons of high-quality water a day by 1987 for its nuclear fuel re-processing plant at Windscale. (The prospect of a planning inquiry highlighted the problem facing those opposing the proposal. They found it extremely difficult to get adequate information and the necessary funds. The Water Authority had earmarked £100,000 for an expected three-month inquiry.)

The water feeds the Windscale plant and the plant contaminates the sea. About a quarter of a tonne of plutonium

Windscale. This nuclear power plant in Cumbria also reprocesses nuclear fuel from Britain and abroad. Sir Martin Ryle, the Astronomer Royal, estimates that the capital cost of a nuclear power station, if spent instead on energy saving would save three times more energy than the station would provide in a lifetime. (Photo: *The Guardian*)

had by 1981 been discharged into the Irish Sea, and the rate continues at about 15 kilogrammes a year. Recent research shows that it is not necessarily the annual amounts of radioactive material that is significant, but the cumulative amount. A study by the Oxford-based Political Ecology Research Group says: "in our view there are ample grounds for believing that major remobilisation of sediment (following a storm, for instance) or more slowly as a result of geological processes, could lead to unacceptably high exposure along a significant section of the north-east Irish Sea coast".

Other countries, including Germany and Japan (which exports nuclear waste to Britain) have reduced discharges to a much greater extent than British companies. No other country, says the Oxford Group, has discharged the products of uranium fission into the environment "so liberally" as Britain. In the US under present legislation, Windscale would not be given a licence.

Some of the most beautiful and valuable stretches of countryside are also under threat from the oil industry: another vested interest to promote development rather than considering ways to save energy costs. It has been granted licences to sink wells in the North Yorkshire Moors, the Peak District, the Cotswolds, the Lincolnshire Wolds and the Forest of Bowland in Lancashire. All of these are designated areas, either of Outstanding Natural Beauty or National Parks.

Shell acknowledged at the public inquiry early in 1982 into its plans to drill for oil in the New Forest — which provoked the biggest-ever opposition to a development scheme in the area — that the deposits would provide just 3½-10 days of consumption. Shell's initiative, backed enthusiastically by the Department of Energy, threatens not only a haven for peace, but a habitat for wildlife including hawks, badgers, and butterflies.

The consortium which discovered the oil at Wytch Farm, Dorset, in 1977 promised to make an effort to protect the countryside and wildlife. But with the growing pressure for more and more energy resources (despite economic recession and waste of existing resources), the search for oil is almost certain to continue threatening the countryside and farmland, especially since the cost of drilling on land is much less than drilling beneath the North Sea. Wytch Farm probably cost the consortium £10 millions; a field of a similar size in the North Sea would have cost about £500 millions to develop. Oil produced in Dorset is worth about £200,000 a day and the are around Poole could prove to be Europe's biggest onshore oilfield. Local farmers there have persistently complained about dry throats and cattle going off their food when excess gas from the oil field is being burnt off.

In 1979, another consortium, Consolidated Oil and Gas (UK) Ltd., based in Calgary in the heart of the Canadian wheat belt, applied for permission to prospect for oil in the picturesque Cotswold village of Guiting Power. The seeds of a familiar debate: will this ruin the character and attraction of this village in England's heartland, or will it save it from lingering death as a rural backwater providing houses for a few old people or second homes for those who normally live in the cities?

The local council was on the company's side and it had the crucial support of Raymond Cochrane, the lord of the manor who owns half the village's 113 houses and 1,000 acres of surrounding land. He has protected the buildings and formed an amenity trust but insisted: "We are trying to keep one Cotswold village where local people can live". Yet will the oil bring any economic benefit to the village?

The Scotland Hydro-Electricity Board wants to prospect for uranium in the Orkneys against heavy local opposition. Companies were offered large grants from the EEC to assist them. With more enthusiastic support from the local population, Consolidated Gold is exploring for uranium in Scotland's Grampian region. The only place where the UK Atomic Energy Authority (UKAEA) faced no opposition against its search for sites to dump nuclear waste was in Caithness, perhaps because most people in the area in question work for the Authority's reactor at Dounreay.

The UKAEA attempts to defend its plans to dump waste on the grounds that any decision would affect future generations, not us. Dumping would not take place for a long time. It delayed naming the sites which its experts were eyeing, yet insisted that they were for trial drilling merely for "basic geological research" and that, in any case, underground nuclear waste tips would cause only a fraction of the damage mining would involve if coal-based energy was used as an alternative.

Scotland, for the AEA, has the advantage of hard granite. But the Authority also looked in England, in the Cheviots, at salt deposits in Cheshire, Somerset and the Hampshire/Dorset border, at clay in the Worcestershire basin and the area known as the Widmerpool Gulf which includes the Vale of Belvoir, where a major battle is being fought between farmers and the National Coal Board (NCB).

Then suddenly in early 1982 the Government decided that radioactive waste may, after all, be kept above ground, indefinitely. It was concerned not only about the high cost of the proposed drilling programme but the fact that the programme had also become a focus for environmentalists and anti-nuclear groups whose support Whitehall remains desperately anxious to diminish.

Ironically, one of the main planks in the landowners' case in Belvoir is that the NCB's plans do not take into account alternative sources of energy, including nuclear. And while the nuclear waste programme is formally under the control of the (weak) Department of the Environment, that Department is also responsible for the public inquiry into the Board's plans to sink three new coal mines in the Belvoir area. "Do they want people who feel its a choice between mining for coal or having atomic waste dumped on their doorstep", asked Chris Tizzard, spokesman for the Vale of Belvoir Protection Group?

After its largest single discovery of coal this century, the NCB wants to mine 7½ million tonnes a year from the late 1980s from three seams under Hose, Asfordby and Saltby. The coalfield, the largest in Europe, is estimated to consist of 500 million tonnes and coal would be mined beneath ninety square miles of land. But Belvoir is also a test case: it is part of a major initiative by the NCB — its 'plan 2000' — for a fifty percent increase in deep-mined coal production, costing £500 million annually with twenty new mines sunk each year.

The battle lines were drawn up with the National Farmers' Union, the Country Landowners' Association, local people, the Duke of Rutland, who is owner of Belvoir Castle and former chairman of Leicestershire county council, the

Leicestershire and Nottinghamshire councils, Melton Mowbray district council and conservation groups on one side. On the other, the National Coal Board, which spent at least £8 millions of public money preparing its case, and the Department of Energy.

The Board insists that it would take up just 3,000 acres of farmland but, on the basis of over-optimistic forecasts in the past, the objectors say it would take up at least four times as much, including a significant amount of Grade 1 farmland. While farmers' groups see the fight in terms of food production versus coal production, many of the anti-development groups are concerned mainly about the prospect of up to 4,000 mineworkers invading this peaceful country area (although Belvoir Vale and Castle attracted 300,000 visitors a year). Michael Latham, the Conservative MP for Melton, wrote in *The Field*: "as the miners come, with their wives and children, pressure would grow to find jobs for their families . . . by the year 2000, the whole area would be completely different and the pressure on the remaining agricultural enterprises of urbanisation would be intense".

More significantly, perhaps, the inquiry again showed up the inconsistencies and the sheer guesswork used in the subjective energy forecasts used by the power industries. Objectors at the Belvoir inquiry, encouraged by the uncertain evidence given by Department of Energy civil servants who were told by Whitehall that they should not discuss the merits of Government policy, questioned whether the coal will be needed, at least in the time-scale suggested by the NCB. And the Central Electricity Generating Board (CEGB) had yet to make any commitment to buy the coal, saying only that it all depends on whether the price was right. Meanwhile, the CEGB itself looks for sites in Cornwall for a nuclear power station even though the Cornish consumption of energy has fallen over the past few years. It does so in spite of the fact that an overwhelming majority of county councillors are opposed to the scheme. Over 90 percent of the population of the parish of Luxulyan objected to the Board's plan to explore the surrounding farms for a suitable site. Yet it went ahead and reneged on earlier promises by bringing in contractors from Yorkshire to do the work.

Limestone Quarry, nr. Matlock. *The search for stone, sand, gravel and clay continues inexorably as pressure on what lies under the land increases.* (Photo: Don McPhee)

The National Coal Board has earned a good reputation for the care it has taken to try and restore land to agricultural use. In Northumberland, for instance, it has returned 2,600 acres to farming though it then proposed a plan for open-cast mining over 20,000 acres. Good quality farmland in Cumbria, Flint, Derbyshire, Staffordshire and Oxfordshire could also be threatened by coal mining while an NCB map shows potential for mining across a broad sweep of South and Central England from Somerset to London.

It is a geological fact of nature that some of the most sought-after minerals used by industry are in some of the most beautiful, and officially-protected, areas of the country.

Rio Tinto-Zinc is still after copper in the Snowdonia National Park. The Peak District National Park planning control committee has allowed Dresser Minerals International to mine fluorspar, an ore widely used in steel, aluminium and chemicals, with the support of a local community, the village of Youlgreave, in search for more jobs. ICI has been allowed to go ahead with the largest limestone quarry in Europe, in the Peak District, although the company (which has a powerful influence in Whitehall) did not have to prove national need and the investment will not lead to any significant increase in employment.

The countryside and the environment are under even greater threat because of the recession. It is a victim of the jobs versus ecology argument cynically perpetuated by companies which have never liked pollution or planning controls. Now, they threaten local authorities and the Government, saying that if you insist on protecting this part of the landscape, force us to adopt clean air or clean water policies, then we will have less money to invest in local jobs.

A dramatic example of how the threat to jobs has influenced a decision by local authorities is the long-running battle between Bedfordshire county council and the London Brick Company. Early in 1981, the company was given the go-ahead to build a new plant without incorporating the pollution controls the council had twice previously demanded. The council's decision to reverse its position came just a month after the company announced over 1,000 redundancies. The Bedfordshire brickworks, which emit both sulphur and

Polluted River. *(Photo: Ken Lambert © Camera Press)*

fluoride, have been closely investigated by the Alkali and Clean Air Inspectorate. But the new chief inspector, Jim Beighton, in his 1980 annual report, appeared to have given in already. "Care and concern for the environment," he said, "are now having to compete more with such factors as inflation, employment and difficult trading conditions". As the Government is encouraging companies and its own officials to turn a blind eye to the Factory Acts and the Health and Safety at Work Act, so it is delaying the implementation of part II of the 1974 Control of Pollution Act which should have been introduced already and is designed to force companies to clear up effluent now being poured into rivers.

The British Steel Corporation continues to pollute the air through cracked and old coke ovens. It is also polluting the River Usk in Gwent. Meanwhile, unions are supporting plans for Steetley Minerals to destroy an SSSI at Thrislington, county Durham, to quarry dolomite which is used for steel production. Without this plan, the firm has warned, over 500 jobs would be at risk.

But the attempt to combat the recession or push for short-term solutions and (from the companies' point of view) quick profits, is rather like walking up a down escalator. Companies are putting off a decision to face constraints which will become inevitable at some stage, if only after more land has been needlessly destroyed. Of three west coast beauty spots in Scotland chosen for the construction of North Sea oil platforms by 1981, Portadie in Argyll had never been used, Ardyne, near Dunoon, had been empty for more than two years, and Loch Kishorn in Wester Ross had produed only one platform.

And, in the meantime, the Government chooses to brush aside ample evidence that shows environmental controls can lead to more jobs and even better profits, let alone better working and living conditions.

The country is being invaded by motorways. About 30,000 acres of farmland were lost to motorways between 1968 and 1978. Each mile of motorway uses about 42 acres of land, the equivalent of thirty-five football pitches. Each mile of road requires some 10,000 tonnes of sand, gravel and hard core — mined from other land elsewhere. Motorways make it much easier for people to see friends in other parts of the country

Dump at Skipton. *(Photo: The Guardian)*

and to enjoy the countryside. They also make villages more accessible to those who work in large towns, and so encourage villages to be filled with weekend second homes and those with little interest in the local community. They also push up the price of houses beyond the reach of the existing local population.

The countryside is under threat from other traffic. The proposal to extend Stansted airport in Essex will lead to the loss of 4,000 acres of prime farmland with the prospect of a further 15,000 acres taken up by related development such as industrial warehouses and accommodation for airport staff.

The most strikingly ironic pressure on the country comes from rubbish and waste. Finningley gravel pits in South Yorkshire, an area officially protected because of its rare water plants, was recently filled in with waste from a maggot farm. Monmouth district council wants to tip rubbish on Great Doward Hill on the Gwent-Herefordshire border — an area of old beechwoods with wild orchids adjoining two nature reserves. The Crymlyn Bog, near Swansea, the largest lowland fen in Wales and a Grade 1 SSSI is under threat from refuse tipping by the city council and the CEGB.

An estimated 17 million tonnes of domestic refuse is produced each year and the bulk of this, about 85 percent, is disposed of on the land, frequently in wetland areas. The amount of refuse could be significantly reduced, of course, if people were encouraged to be more aware of the ways and the value of wasting less or to return bottles and paper for recycling. (It may seem surprising that farmers have offered their land for tipping domestic and industrial waste. Members of the Country Landowners' Association in Shropshire have encouraged the use of former 'totally useless' peat bog near Shrewsbury to be filled in with local council waste. The idea is to cover the rubbish with clay and then spread the original peat, which has been kept to one side, on top. In another farm nearby, a former deep valley has been transformed into flat pastureland used for grazing cattle and sheep as a result of tipping). But local authorities are now looking at green field sites to get rid of their waste. Land in Britain is also being used as a dump for toxic and dangerous chemical waste, and although legislation designed to control tipping was belatedly introduced, it is frequently ignored and in a typically secretive way information about what is happening is kept from those who are most directly affected — local residents.

One notorious case concerns the dumping of dioxin after an explosion at the Coalite factory near Bolsover, Derbyshire, in 1968. Without the knowledge of most local councillors and planning authorities — and those who did know were told to keep the information to themselves — the company dumped toxic waste in an area of disused open-cast coal mines near the villages of Morton, Stretton, Shirland and Higham. Few people know exactly where, and those who do refuse to say. The Severn-Trent Water Authority has given bland assurances to local inhabitants while Derby County Council said in a refrain reminiscent of Whitehall: "It is not in the public interest for the location of the deposit to be revealed". Another large toxic waste tip has been licenced next to where the dioxin dump is believed to be. The area is close to a hospital and a housing estate. Drums of chemicals have not been buried and lie on top of the soil. Contaminated water is pumped into local streams where children play, and sometimes swim. And a private company was given permission to mine for coal, though it was not told about the dioxin dumped in the area, and animals have died without proper post-mortems being carried out.

Inhabitants of the village of Woodham, Berkshire, have persistently complained about the smell and other symptoms — including headaches and skin rashes — caused by a toxic chemical tip which is also used by the County Council as a dump for domestic waste. The Thames Water Authority in 1975 warned the operators to stop pumping contaminated liquid into a nearby stream. The county surveyor acknowledged that planning regulations were being broken. The company, in 1981, told a House of Lords committee on science and technology investigating the whole question of toxic waste: "The site, although at times not entirely free from smell, is in our opinion no more of problem than the adjacent chicken and pig farms and nearby brickworks".

Legislation, including the 1974 Control of Pollution Act, is woefully inadequate and, even so, consistently ignored.

Under Government proposals announced in 1981 controls

over the disposal and removal of toxic chemicals were taken out of the hands of local authorities. Although a licensing and registration system exists, how effective it is will depend ultimately on the honesty of the producer or contractor or the alertness of local administrators.

Awareness in Britain of the dangers involved in dangerous waste is nowhere near the extent expressed in other Western countries. Those whose job it is to ensure that controls are respected, mainly local authorities, prefer, in the traditional British way, to have informal chats with those companies responsible even though all the evidence suggests that this emphasis on pragmatic persuasion rather than statutory controls, simply does not work.

Financial pressures on local authorities have led to a cutback in the number of inspectors — the Greater London Council in 1981 had no toxicologist on its staff — and Whitehall has abandoned the former notification procedures on the grounds that since toxic waste dumps have to be licensed, notification about the movement of waste is unnecessary. While some authorities are genuinely concerned, others have welcomed waste (and the dubious financial rewards — it is the tip owner who amasses the profits). Waste from as far afield as Cornwall and Scotland is transported to Essex, which has the privilege of harbouring the country's largest tip in Pitsea.

Even the Department of Environment overruled Devon County Council in 1980 when the Council tried to prevent a tip owner from dumping millions of gallons of waste in a deep pit at Higher Kiln, listed by the Institute of Geological Sciences as 1 of 50 tips most likely to pollute water supplies.

Council homes are still being built on old dumps, tips, gas works and chemical factories. There are potentially dangerous asbestos dumps — there are no records of precisely how many — in Hebden Bridge where 70 local inhabitants have already died of diseases caused by asbestos poisoning. An underground fire has been raging for over 3 years in a tip north of Sheffield.

Not content with adopting a complacent and thoroughly irresponsible attitude to dumping its waste, Britain is encouraging imports of other countries' toxic and dangerous substances on the entirely spurious and cynical grounds that it is to our commercial advantage — so Britain has earned the reputation of being a 'toxic dustbin'. British Nuclear Fuels has gratefully accepted nuclear waste from Japan for reprocessing at its Windscale plant. A British-based firm, Riafield, in 1980 signed a contract with a Dutch company to import phenol waste, which are highly toxic chemicals. The Dutch had just learnt that a housing estate in Lekkerkerk, near Rotterdam, had been built on top of a chemical dump. Chemicals had corroded water pipes and the water supplies were being contaminated. Thousands of tonnes of the phenol-carbolic acid ended up in storage tanks near Southampton. Other tanks with Dutch waste were directed to the Thames estuary and Humberside, where toxic waste was being piled up at the Immingham Storage Company. Part of the waste was there simply because the ship carrying the chemicals was said by the captain to be in danger of breaking up in the North Sea. A spokesman for the company said: "I don't know what all the fuss is about. You could go swimming in this stuff without any harm".

British Petroleum has allowed companies to use its cleansing facilities for tankers on the Isle of Grain in Kent only to discover later that phenol waste, not just oily water, was being unloaded. Companies eager to make a quick profit (Riafield had told Essex county council that waste it wanted to land there would eventually be re-exported to South Korea where an importer wanted to buy it as a tax loss to enable him to bank western currency in Europe) quickly latched onto a loophole in the 1974 Control of Pollution Act which inadvertently repealed an earlier law whereby the storage as well as disposal of poisonous wastes had to be notified to local authorities.

And not so long ago, a development plan for part of Knightsbridge in London was rejected on the grounds that there are plague pits under the land and that they might have caused another outbreak if they were exposed. The plague virus takes a thousand years to die.

10

RURAL DEPRIVATION
AND COUNTRY ATTITUDES

"Ill fares the land, to hastening ills a prey,
  Where wealth accumulates, and men decay.
  . . . Trade's unfeeling train
  Usurp the land, and dispossess the swain;
  Along the lawn, where scatter'd hamlets rose,
  Unwieldy wealth and cumbrous pomp repose."
– *from* The Deserted Village, *by Oliver Goldsmith.*

Farmland owners (if we can omit here small farmers in poorer regions) can be divided into two distinct groups. There are the established landowners who have inherited their property. They, for the most part, would like to preserve the status quo ante; they share an instinctive suspicion of development or new agricultural methods, and are sensitive to the need to protect the countryside. They are generally sympathetic to those who point to the gradual destruction of woodland, hedgerows, national parks, moorland.

Then there are the farmer-businessmen, those who have recently increased their holdings; some of them are career farm managers working for large institutions. These are determined to invest and benefit as much as possible from available grants and subsidies. They closely watch shifts in the EEC agricultural pricing policy. They are intolerant and impatient about the cries of alarm from conservationists and others who warn that things cannot go on as they are for the sake of the long-term good of the community or of the soil. They are a dangerous, an increasingly powerful breed, with a short-term outlook. They are confident that, as other sectors of the economy struggle in the recession, their future is certain. They are also privileged, and encouraged, perhaps above all, by their monopoly of the labour market, and by the reality that in many country areas the closed shop benefits the employers, not the employees, the workers or the unions.

And this is where the two groups do have a common interest; they do not want to relinquish control over the country. While the latter can rely on labour and productivity — economic control over the countryside — the former have traditionally dominated parish, local, and even county councils. A feudal attitude remains; sometimes it may be a relatively enlightened and liberal, Whiggish, variety; frequently it is more intolerant and with brutal consequences. It is always presumptious, carrying with it resentment against 'outside interference' or the independent views of a minority.

The wife of a tenant farmer near Basingstoke says: "The rich and greedy have discovered there is money in farming, and have moved in to exploit the potential. We now have land-owners taking in farms, employing managers, and farming units of thousands of acres, as do the City institutions. Farming on this enormous scale is successful only by using vast quantities of poisonous and expensive sprays to combat the ever-increasing and resistant cereal diseases and weeds. Profitability on the animal side is achieved by keeping large numbers in unnatural conditions, continually dosed and injected with drugs.

"My family have farmed here since the 1880s, my father being the last tenant. However, for the last 20 years my husband has farmed here, while my father lived quietly in retirement, assured by our landlord we could carry on the tenancy. On his death in 1973 our landlord informed us that he would give us a 'partnership' in place of a tenancy, because of tax. Initially we agreed, expecting a fair agreement. But after waiting a year we were given the following terms: in return for the use of the land, we paid a sum equivalent to rent, plus 10 percent of profits, responsibility for insurance, repair, and replacement of all buildings, many of which are near derelict, and only six months security of tenure. Everyone we consulted in the agricultural profession said it was unacceptable, but our landlord said we either accepted it or give up the farm.

"We decided we would not vacate the farm, and our landlord has put in a writ for possession in the High Court.

We feel sustained in our decision, not only by feeling we are fighting injustice and greed, but also that hopefully we shall keep one more farm tenanted, and it will not be added to the 2,000 acres already farmed by our landlord".[1]

That is an example of what is happening, and it is by no means an extreme one. Meanwhile the squire of Great Barrington, near Burford, owns his ancestral home, Barrington Park, and the entire village. Parts of the village, including listed cottages, are falling to bits. But the tenants live rent-free and praise their landlord. Cotswold District Council is threatening a compulsory purchase order because of the state of the houses.

That is an example of another kind of landowner. Yet the landed interest, its old and new ingredients, does not want to relinquish the control it enjoys. Large landowners in the more thinly populated areas of Britain, and Suffolk is a good example, have consistently blocked proposals to promote light industry or housing schemes which may become focal points opposing the status quo and loosening the landowners' grip over the housing market.

Every group and organisation represents a vested interest, and there are many which claim to represent the interests of all the community. There are many individual examples. In Hampshire, for instance, Sir John Scott, recently appointed high sheriff, stands as an 'Independent' (though he used to stand as a Conservative, and no Conservative stands against him now) in county elections, He is a large local landowner, a Justice of the Peace and former senior local official of the Country Landowners' Association. He presumes to win, in the interests of the whole community, and does so.

Even the National Trust can be presumptious — not as a caretaker for the nation of stately homes, but as a landowner. In 1977, James Hughes applied for planning permission to convert the back of Avebury Chapel into a combined bookshop and post office. The idea attracted the full support of the pastor and of the vast majority of the local population. But it has been consistently opposed by the local landowners and the National Trust, the main landlord in the area, whose own bookshop was set up after the village had been listed as a potential conservation area. Another bookshop, owned by the

Department of the Environment, was opened after it had been designated an official conservation area. And the village already has a small grocery-tourist shop, and there is a kiosk at Avebury Manor itself.

The parish council voted down the Hughes plan by 3 to 2. Its decision was upheld by Kennet District Council and later by a Department of Environment inspector, a local architect and a member of the National Trust. Among the objections was the claim that the shop would lead to the display of "an inappropriate number of signs" and to "the crossing of the A361 to the detriment of the free flow of traffic and of highway safety". The village remains without a post office, a victim of arbitrary attitudes.

In 1944, Great Tew in Oxfordshire was proclaimed "the prettiest village in England". Thirty years later its listed seventeenth century cottages were falling into ruin. They are owned by the local landlord, Major Robb, a distant and old-fashioned character who owns 14 farms on 5,000 acres of good quality land. His agent and partner, James Johnston, is a leading light in the National Farmers' Union[2] They have resisted attempts to improve the state of the village, though the Department of Environment eventually offered grants to preserve some of the houses.

These are just two further examples of how villages are suffering and how the pretty, calendar-photo, image belies the reality and indeed compounds the problems. Rural areas have been neglected: analyses of poverty in Britain have been biased towards urban areas. And despite — or because of — the popularity of television programmes on country life, including such programmes as 'The Archers', the mass media maintains that urban bias.

Yet the lack of mobility and local opportunities, and the routine nature of much of rural life means that the rural population watches television at least as regularly as the urban and suburban population. Local community entertainment has largely disappeared, making rural living even more lonely. The more urban we have become and the more urban the attitude and philosophy imposed on us, the blinder we are to the opportunities and advantages of small, local communities. But these communities in rural areas have also been forced to

Widdicombe Village, Devon: *This was the home of Uncle Tom Cobbleigh and all. It has calendar attractiveness. The other face of rural life is poverty, deprivation, and a decline in public services. (Photo: Camera Press)*

lose their self-reliance, with services cut and more distant, with help taking much longer to get, with local authorities (and Strathclyde in Scotland is one of the worst examples) larger, further away, and more centralised.

Beyond the motorways, behind the stately homes and the second homes, rural life suffers from deprivation and poverty.

Many local authorities use caravans as a way of solving housing problems. The Isle of Wight Council 'allows' people to live in caravans out of season though they are not licenced as being fit for dwellings in the winter months. In the New Forest, families spend the winter in caravans and are then told to get out when the holiday season comes round. Every time they move, they lose their place on the housing lists. Poole Council in Dorset dumps many homeless on a caravan site on Rockley Sands. The poor are frequently exploited when they apply for sites since there is no alternative accommodation. Andrew Larkin described a tour of Dorset[8]: "It provided practical evidence that the poor who are badly housed in rural areas often suffer greater hardship than their urban fellow-workers. There were young families living with their parents who had been on the housing waiting list for about ten years, whilst their children grew up in more and more cramped surroundings. There were those who had ended up walking the streets with their children as a result of brutal homelessness policies.. . There were families living in rural slums, isolated privately-rented cottages with no basic amenities which had suffered years of neglect and would be rented out until the point was reached when even the most desperate would not take them, at which point they would be left to decay and fall down. Such cases, although they may be isolated examples in any particular village or rural town, build up into a picture of neglect of rural housing problems which has lasted for decades".

Rural authorities often allow local councillors to nominate tenants for council houses in their parish. Howard Newby has described how the mid-Suffolk Council, including its few Labour councillors, criticise what they say is the 'bureaucratic points system' for allocating council houses preferring, instead, recommendations by the councillor responsible for the ward where the house is situated. The council does not want to be tied down by any policy or encourage the development of "a class collective consciousness among the unprivileged". This is a justification, of course, to allow them to rely on personal decisions and personal control, with all the subjectivity, discretion and unaccountability this involves.

It is one manifestation of modern feudalism, still strong, though more sophisticated and more disguised than in the past. In most rural areas, for example, the magistracy includes the large local landowners. About 30 percent of the agricultural acreage in Hampshire is owned by people connected with local government. In Suffolk, virtually all the key posts in the County Council have been in the hands of large landowners including the chairman and vice-chairman of the planning, education, social services, finance and policy committees.

The Labour administration in 1976 abolished the tied cottage system — in theory. The legislation went ahead in spite of sharp opposition from the farmers' organisations which argued that the system was essential to ensure that workers lived close at hand, especially on livestock farms.

In spite of the steady stream of evictions, and the continual threat of many more, farmers have tended to use the tied cottage as a carrot as well as a stick. Many farmworkers defend the system since it gives them a kind of easy, superficial, security; they have been lulled into accepting, even enjoying, total dependence on the landlord. They have learnt even to accept that the landowner has the right to interfere in their personal and family life. But one of the main advantages for landowners, and the landed interests in local authorities, is that the cottages secure a captive labour force and provide another excuse to restrict local expenditure — and, therefore, rates — and limit the building of alternative, low-cost accommodation. There are long council waiting lists in rural areas, demonstrating that the demand for cheap, rented accommodation, remains high.

In their 1966 Survey of the Dedham Vale, the Suffolk and Essex planning authorities said: "In our view, the traditional peace and tranquility of the Vale depends on keeping the population scale more or less the same as it is today. Natural increases from within the Vale should not necessarily be housed within its villages . . . No further private or public estate development will be permitted." Fewer than three new council houses had been built in the Vale in any year since the war. And the prejudice against estates, and those who live in them, a prejudice that is more suppressed and diffused in cities, is much more evident in rural areas. "They must come from the estate", or "there's an estate in that village, you know", are common remarks that greet a crowd of unemployed youths on street corners, or a theft of a bicycle or graffiti in the bus shelter.

The pressure on rural housing is compounded because of the double demand — from local residents and from better-off commuters, second-home owners or retired people from the towns. Local councils and traditional communities should limit planning permission and force developers to subscribe to social and more human priorities. And Britain should follow the practice in other countries where limits have been introduced on second-home ownership. Indeed, local authorities in Wales have already tried to use section (52) of the 1971 Town and Country Plannning Act to restrict the use of existing houses. An attempt by Gwynedd County Council to do just this to two landowners in Anglesey was bitterly opposed by the Country Landowners' Association as being "the thin end of a very nasty wedge".

In spite of the consistent prosperity, and high productivity record of agriculture over the past twenty years, farmworkers remain at the bottom of the wages table along with sectors of the food and catering trades. One in six of all Family Income Supplement claimants are farmworkers – the largest single group of claimants. Agricultural wages are particularly low in East Anglia, the richest farming area in Britain but an area with few other employment possibilities. In the mid-1950s a farmworker with two young children did not pay tax until he was earning slightly above the national average wage; today the tax threshold is less than half the national average wage[4]. In 1979 nearly half of all full-time farmworkers were low paid, according to official criteria. They suffer from the poverty trap, from the higher cost of services in rural areas. Farmers can also keep down wages by resorting to casual labour and get round the Equal Pay Act by the same device. In 1977-78 when the average net income of larger farmers — those who employ hired labour — stood at £20,500, farmworkers paid, proportionately, twice as much tax as farmers. The gap between agricultural wages and those in industry is as wide as ever.

Country Services

The decline and neglect of country services has been well documented[5]. Small communities in rural Britain are involved in a struggle for their survival as more and more of their essential services are closed and public transport links deteriorate, making it steadily more difficult for the elderly and less affluent countrydwellers to continue living in these villages.

Thirteen percent of the villages in Gloucestershire and Wiltshire lost their village shops between 1972 and 1977. Ten percent of all villages in Cornwall had lost their post office in the nine years up to 1976, and nineteen percent had lost their doctor's surgeries over the same period.

More than 80 percent of the total village population have to travel more than 2 miles to a chemist, and more than half the villages in Britain are without a doctor. Most rural areas are classified as 'restricted' by the Department of Health and Social Security, and it is impossible for a new doctor to set up new practices in these areas.

In West Dorset, half the population live in villages of less than 500 inhabitants; 75 percent of the villages have no school, nearly 70 percent no garage, 60 percent no pub, half no sub-post office, and 30 percent no shop.

The increasing centralisation of services, including transport, assumes that people have their own means of access to these services, and notably their own cars. Yet more than half the pensioners living in Norfolk, for instance, have no car. Ten percent of Devon villages which had a mobile shop in 1967 had lost it by 1975. Yet the Department of Environment concluded in a survey of the County at that time by stating: "Clearly people have adapted to the absence of a bus: those who could not, presumably left some time ago". Rural bus travel has been cut by 50 percent over the last twenty years. The price of petrol is higher in the country than in traffic-congested towns, and petrol companies are threatening to close more country garages arguing that it costs too much to distribute petrol to outlying areas.

Herefordshire has been described by its own local authority (part of the new 'County' of Hereford and Worcester) as "the geriatric ward of the Midlands" and that in fifteen years it could be turned into a human desert. Ninety percent of the people living there do not know which authority was responsible for their services and eighty percent of those in the more remote areas do not know they had a local councillor. The survey in 1978 reported a growing lack of services and facilities, high costs of heating and food. Thirty-five percent of all families in the area had no car.

A survey of the village of Trent in Dorset showed that there was an increasing number of elderly inhabitants, and fewer young. More people had lived there for less than a year than for any other period of time. Over half the population commuted to work. Most new residents earned more than £6,000 a year, while most of the 'locals' earned less than £4,000.

As villages nearer towns have been actively populated, so the polarisation between country-based poor workers and employees and the richer town-based ones, increases. Village schools are fading away: 800 were closed during the decade up to 1977 and a few years later were closing at the rate of more than one a week. Somerset council closed 154 schools out of a total of 500 between 1947 and 1974. It then wanted to spend £3 millions on new school buildings. As parents and teachers up and down the country never tire of pointing out, village schools and school halls can be a powerful focus for rural communities. In Scandinavia, experience of large-scale rural school closures in the 1950s and 1960s led to a drastic reappraisal and many schools are being re-opened. Sheila Browne, head of the Schools Inspectorate, admitted after a survey into primary education was completed in 1977: "No doubt I shall seem out of tune with trends and the state of the economy if I pull out only one point now, namely that (the Inspectorate) has the impression that rural schools — and I am not saying small rural schools — may be in some respects nearer the heart of the educational matter. I quote one of the staff inspectors for primary education: 'In rural schools, children had more opportunities for informal talk, and hearing and reading good stories and poetry. They had more opportunities to select books and turn to reading for pleasure. Most children had more challenging tasks in mathematics . . . In science they were more likely to make effective use of the environment and were generally better provided for. Greater emphasis was laid on the acquisition of moral and ethical rules and values' ".

The Standing Conference of Rural Community Councils estimated in 1978 that the cost of educating a primary school child in the 12 most rural counties is only £3 more than the national average of £321. A small price to pay for the advantages claimed by the Schools Inspectorate (though do the figures simply reflect the determination of rural authorities to keep expenditure — mainly rates — down to a minimum?).

In a study of the movement in Norfolk which sprang up to fight the imminent closure of 50 village schools in the county, the National Union of Agricultural and Allied Workers argued that it is not the drift of population away from the land and the villages into towns and cities which led to school closures. The closures were a cause, not the result, of the drift. In 1977, the village school of Gissing had 23 children, one full-time teacher and just about every one of the village's two hundred residents wanted it to stay open. So did the neighbouring village of Burston where the already-overcrowded school would have to accommodate the Gissing children if the County Council's plan went through. Tom Potter, one of the leading protestors, said: "Gissing's a dying village. The rectory looks like being sold; the church hall may go up for sale; the rector isn't being replaced; the local council has been refused planning permission for six new houses; and now they want to take our school away", Who would want to live there if the village school went? Possibly old people. It is a viscious circle. The authorities say that the schools will lose pupils because parents no longer want small schools with few facilities for their children, while teachers will be hard to recruit since small schools cannot offer much prospect for promotion.

Few authorities bother to ask questions before they make assumptions. The 1967 Plowden report recommended that schools should have a minimum of 3 teachers and 50 pupils. But rural schools are being closed in a haphazard manner and

Derelect quarry village of Nantgwyrtheyrn, Wales. Once inhabited by low-paid workers employed in dangerous and unhealthy conditions, houses are now offered for sale at knock-down prices. Farm labourers' cottages are offered as attractive second homes for urban-based visitors.

simply as a result of immediate financial problems. There is a growing consensus that village schools with their intimacy, sense of community and better teacher/pupil ratio are very good places to learn, and certainly for children up to the age of eleven or so. One solution is to form "cluster schools" uniting a group of small schools, with children (but mainly teachers) travelling to different parts for different activities. Another is to use schools for a variety of activities, encouraging them to become a community centre, providing facilities in the evenings and weekends and a base for adult education. A similar system or network could be set up in urban areas.

In some villages the population declines precisely because there are so few services, and the decline is hastened by decisions of planners who prevent further development precisely because of the lack of services. The village of Waltham St Lawrence with Shurlock Row, wrote Mrs Joan Yeo Marsh to *The Times* in 1978, was 30 miles from London with easy access for commuters. It had retained its agricultural character. "The current concern for conservation in this case threatens to turn the village into a commuters' roost", she added. "The following planning requests are among those rejected by officialdom even before 1974:

"1: A carpenter; to replace his unsightly workshop fronting on the village street by an appropriate house, set back to the building line of adjacent houses, with a workshop at the back, in no way disturbing the amenity of the village.

"2: A teacher of art who is also a potter, to make a small workshop in his garden, and place a sign, in good lettering, at his gate, to advertise his work.

"3: A housewife: to trade in antiques from her house on the fringe of the village.

"Surely", Mrs Marsh concluded, "such activities as these are what a village needs. Can there not, even within the Green Belts, be discrimination between that which belongs to village life and 'industry' which could destroy it?"

Motorways bring accessibility and the commuter zones are expanding: Somerset villages are within what is now regarded as acceptable daily driving distances to Bristol; Birmingham is eminently commutable from the rural heartlands of Gloucestershire and Staffordshire. The Cotswolds are crammed with second homes. In Cheshire, run-down cottages are selling for £35,000 or more and are then renovated with the help of Government grants. An advertisement in *The Sunday Times* in 1978 appealed to the market: "Best of all," it said, "if you want something really cheap and don't insist on a sylvan glade, is to follow the latest trend and look in some of the semi-derelict industrial villages in South Yorkshire. Here you might pick up a two-up, two-down in a terrace for £3,000 or so. You have the advantages of accessibility and mains services. And if you want romance, well, there's a certain poetry about the Industrial Revolution, isn't there?"

Against the background of this kind of effete approach, rural areas are being taken up by second-homers. There are more than 30,000 second homes in Wales and in Gwynedd, there are twice as many holiday homes as there are families on the council waiting list. In some villages there, over forty percent of the homes are holiday houses, empty for most of the year.

Meanwhile, the polarisation between town and country continues apace. When the village schoolchildren visited the farm of Brian Hughes in Essex, only 3 out of 40 had any connection with the land, and two of those were from the same family. There is a fairly ordinary local farming club in the south of England, in the Home Counties, with thirty-nine members, of whom 20 are over sixty years' old, and ten of those are over seventy.

There is a view among planners and bureaucrats that villages are an obsolete settlement form whose continuing existence depends on the forces of inertia in a predominantly urbanised society. Villages, they add, are also archaic in the economic sense. Experts and planners pigeon-hole types of villages under the titles 'working', 'retirement', 'dormitory', and even 'mixed'.

Planners suggest that farming and related trades should employ just 25 people a square mile. They would be supported by centres providing services, with a threshold of 3-5,000 people. County councils have attempted to 'rationalise' rural development, concentrating on 'key settlements' totally unrelated to most old market towns and frequently involving housing estates dumped on the edge of existing towns. It is all part of an apparently inexorable process of bureaucratisation, centralisation and unaccountability that has been steadily reducing the role of the individual and increasing the power of the State over the past 150 years. Rural population declined, at first dramatically during the advent of the Industrial evolution, and then more steadily though no less widespread: eighty percent of Norfolk villages were still losing people in 1979. The capitalisation, mechanisation and intensification of agriculture sped the process along, cutting down the number of farmworkers. On present trends, there will be no farmworkers left in Devon by the year 1991. The number of farmworkers has probably reached the point when, if many more leave the land, even the most capital-intensive of farmers will be in trouble; though as I write farmers are installing computers and encouraged to consider driver-less tractors.

While fewer and fewer people are working on the land, more people are leaving the large cities and towns to live in the country. The interim report on the 1981 census revealed a dramatic shift out of London, and gains in East Anglia, the East Midlands, the South-West and Wales. These movements are taking place against the background of a very small increase in the total population of Britain. Whitehall and planners look upon Leicester, Derby, Nottingham, Norwich and Ipswich, Bristol and Gloucester, and South Devon as regional urban growth centres.

We will have to cope with a completely new pattern for which we are unprepared. Against the background of an overall, if only temporary, fall in the school population, East Anglia's 5-15 age group, for instance, is expected to increase by as much as 12 percent. And as the fall in the population of inner cities continues, the problem in many places by 1990 could well be a shortage of labour, even too many homes and too few people with local authorities having virtually to bribe people to live in their less attractive estates simply to preserve them from greater vandalism and worse decay, and the fate of even emptier shells. We have been left with an appalling state of affairs with both inner cities and rural areas neglected. There are new growth centres for light industries, often like food processing, related to agriculture, that are emerging. But they are the result of short-term and desperate attempts to seek profits or provide employment. There is little or no consideration about the future of the land, landownership, or of the shape of local communities.

A range of piecemeal initiatives are being taken in rural areas; small schools are being saved with local ratepayers taking matters into their own hands. Groups of small industries and crafts are being set up, sometimes with the help of the Council for Small Industries in Rural Areas, a subsidiary of the neglected Development Commission. There is room for many more — more use of joint services such as the combination of buses with postal services and milk distribution, bulk-buy groups, consumer cooperatives, mobile health clinics, housing cooperatives, car-sharing services, community newspapers and rural radio services. The Labour Party has suggested a system of 'community service officers' centred on core units in certain villages or small towns.[6]

The totally artificial and bureaucratic boundaries between different departments of local authorities and the difficulties in swapping money around from one part of the budget to another must go. The seeds of appropriate constitutional machinery exists: there are about 7,200 parish councils in England and each one is obliged already to hold meetings of their council open to the public at least four times a year. Parish councils can acquire land "for the benefit of the local community", have the authority to spend the product of a 2p rate (several hundreds of pounds worth even for small villages) and provide amenities.

But all these are palliatives. A revolution in attitudes, as well as a dramatic upheaval in the economic and political structures are needed. We shall return to this, and consider first the prevailing attitudes in the country now.

Country landowners firmly believe that they will prove to be one of the last bastions of individual liberty. They also insist that they are stewards of the land, holding it 'in trust' for the rest of the population and for future generations. They often reflect a tradition of presumptious aloofness, an extraordinarily sharp resentment against those outside who offer their views and defend their own interests. "Farmers and foresters should get on with their job and they will do their best for the landscape which they know and love", says Prince Philip, and farmers applaud. And they erect barbed wire fences and notices warning: 'Keep Out'.

Viscount Arbuthnott, former president of the Scottish Landowners' Federation, puts it this way: "If a family is still domiciled on and managing the lands recorded as having been obtained by their ancestors between eight and nine centuries ago, and if this has been achieved without the benefit of political favour or the profits from any external commercial enterprise, it indicates not only good fortune but also a certain ability to perpetuate and use the natural resources of the land with which the family was originally endowed".

This, of course, begs a whole host of questions, not least the effectiveness of persistent lobbying by landowners' interest groups, an activity which landowners consider to be 'apolitical'. And since land above all resources historically has been either a gift (with titles) or purchased by wealthy industrialists, if the Viscount is to be taken literally, then he is describing very few landowners indeed.

A much more fundamental point about the attitude among many farmers and landowners is made by Professor Donald Denman. He told the first annual conference of the Small-holders' Association at Reading in 1980: "Modern thinking is collectivist . . . one of the blemishes of modern society, in the corporate cartels of capitalism and in the syndicates of socialism, is the lost identity of the work of men". The small farmer, he said, is "the holder of the landright that makes him what he is".

Now he is talking about a declining breed of small farmers who are threatened by the development of large agricultural companies, by the Common Market and by intensive farming methods. They are the farmers, notably in hill areas and in the West, who can legitimately say that the ownership of land is fundamentally different from the ownership of cars and consumer goods. They are close to the land, they work long hours, they do not necessarily want to spend their money on the same kind of things urban dwellers do. Yet, whether they like it or not, they have no access to the ultra-sophisticated kind of entertainment available to others in the towns.

Yet landowners and farmers, small and large, do share a kind of supreme assurance, a kind of confidence that sets them apart from the rest of us and consciously encourages them to do so. The ultimate and ugliest expression of this sentiment was delivered to approving members of the Farmers' Club in London in 1978 when John Vlielander, a prosperous Dutch landowner, said: "I feel that people in farming in the long run will be better off than those not in farming . . . A period of unrest in many countries, or even of war, lies in front of us. In times of insecurity, farming has always thrived".

A number of branches of the Country Landowners' Association have debated the possible results of a nuclear attack on Britain, concluding that farmers will have a special role to play in the aftermath. They have described in detail, as have such farming papers as *Big Farm Weekly*, what kind of food would be available and how relatively uncontaminated it would be, assuming all the time that rural areas will be largely protected.

Yet, just as they see themselves as performing a vital role for the community — and rely on, and demand, increasing subsidies and grants to do so — farmlandowners claim that they do not want to have anything to do with the State. They pretend that they are self-reliant. Norfolk farmers were busy installing their own generators during the last Labour administration "so that they should not be beholden to anybody even for their electricity".[7]

The attitude of country landowners is curiously paradoxical: fight controls and restrictions imposed by Whitehall, they say, but grant the grants and preserve what David Rose of Essex University describes as the 'non-politics of the status quo'. At the same time, fight against an interfering Government with all the traditional political, social and economic weapons at their diposal. The Country Landowners' Association, we have seen, is a highly effective interest group, especially on matters of tax policy. Country landowners do form a very political, closely-knit and successful lobby, whether the lobbying takes place in Whitehall or in Westminster corridors or during country shoots. They do not fight shy of broad political judgements. "The ownership of property is the bulwark against tyranny", says James Douglas, the CLA's director-general. Major Basil Heaton, chairman of the CLA's Clywd branch wrote in a recent annual report: "The prime role of the countryside is the production of food and timber. But the countryside", he added, "suffers from too much decision-making by central and local government and this removes the initiative from individual landowners who, in the end, have the final responsibility for the countryside".

A group of sociologists from Essex University[8] questioned half the farmers on holdings of more than 1,000 acres in East Anglia and a third of the full-time farmers on 44 parishes in Suffolk. They found that seventy-five percent of those with more than 1,000 acres, and over 80 percent of the others, believed that people in general were hostile to them. That is an attitude that reflects the insularity and siege mentality of country landowners, something which their leaders are now concerned about and so encourage representatives of urban local authorities to visit farms.

What makes it so difficult for country landowners to appreciate the attitudes of the rest of us is their own clear conviction that they are a force for good. In defence of the squirearchy, Lt. Colonel Reeve of Leadenham, Lincolnshire, a village of one shop, two pubs, a church and a garage, told *The Guardian* in 1978: "Why are we here? Ten years' ago, we'd have been better off if we'd sold up and gone to live in the south of France. But when they've done away with the rector and amalgamated schools of less than fifty pupils so that the schoolteacher goes too, and liquidated the squire, who will there be left to talk to?" Reeve does not own the entire village — some council houses have been built — but he does own many of the houses, the village shop, the larger of the two pubs, an eighteenth century manor house and 3,500 acres around it. The family owned 8,000 acres when he was a child (the Reeve family has been at Leadenham since 1792 and the female side of the family since the 16th century). He farms 250 acres himself; the remainder is let. If he sold out, he says it would be to an insurance company or a pension fund. The colonel was chairman of the parish council for twenty-five years, a magistrate for 25 years, including ten as chairman of the bench, (almost an inherited job, he acknowledges) and chairman of the Grantham Conservative Association.

Villagers go up to the manor house for firework displays; he provides a house for a blacksmith; he has put down drains and repaired houses. He regards himself as a buffer between the villagers and the whims of the planners (who chopped down the King George Jubilee sycamore tree without consulting anyone). He has fought to keep open the village post office and to save the village fishpond. Institutions, he insists, would not worry about such things.

Roger Paul was recently president of the CLA. He is chairman of Pauls and Whites, an important animal feed manufacturer, a large pig producer, and farmer of 845 acres in Suffolk in partnership with his family. In 1980 he was also a member of the Eastern Region Planning Council, and Chairman of the local Conservative constituency party at Eye. His wife is President of Ipswich football club, the only woman in such a post in the entire football league. He, too, emphasises the importance of what he calls the squire's 'social responsibility'.

It is all very well, he says, for financial institutions to insist that they appoint good farm agents; but they are not there during the weekends, they do not contribute to village life. They are merely doing a job, much like any other, and probably live forty miles away.

A fair point. But many landowners believe that they and only they should be the contributors to village life. An extensive survey found that farmers resented "the invasion of villages by middle-class newcomers", who initially at least are hostile to farmers. And to a mention of the large cars they drive, farmers respond that in spite of the loss of 1 million acres to urban and other uses, farming had doubled production, between 1960 and 1974, of wheat and barley.

Farmers who campaigned against the National Coal Board's plans to dig pits in the Vale of Belvoir, warned of the "ill-mannered and ill-disciplined (miners and families) who keep greyhounds and whippets". A slightly more sophisticated point is made by Roger Paul. People are moving to the country from London, he says, but they do not always fit in. They were the first to grumble about the smell of pigs or traffic visiting farm parks.

And there is an uglier side to the landowners' attitude. They patronise labour and are encouraged to do so by the deference of farmworkers who, as Newby puts it, "subscribe to values which endorse their subordinate position in society". The campaign by the National Union of Agricultural and Allied Workers for higher and more regular wages, threatens to destroy the traditional good relations with their employers", according to Jack McLean, a Gloucestershire farmer. "If we have to reward our workers by order, a lot of things which they get just because we like them are going to go by the board".

"Contemporaries of mine", David Richardson, a prosperous Essex farmer wrote in *Big Farm Weekly*, "are simply not prepared to tolerate what they see as an erosion of their rights as employers . . . I suspect they will not be the only ones to decide that their loss of freedom to hire and fire is unacceptable . . They much prefer to make their farming fit the labour on which they can totally rely and if that means cutting back to what they and their families can handle, so be it". That was in reference to the Labour Government's Employment Protection Act.

Most landowners do not officially recognise the union which now represents only about half of all full-time farmworkers. "Farmworkers," according to the prevailing attitude in East Anglia[10], "take pride and satisfaction in their job despite the weary toil it sometimes entails". They are not unionised because they are "closer to economic realities than their brothers". Farmworkers, it is suggested, succumb to the highly structured and status-conscious nature of rural society, and to the temptation to regard the landowner and employer as the provider of social as well as economic needs and support because the opportunities and public services open to them are so limited.

The attitude of farmers is recorded by an Essex University research team:[11] A farmer on 200 acres in Suffolk was concerned about an apparent deterioration in labour relations which he put down to "unionisation caused by low pay and television . . . This television is a hell of a thing. Ninety-nine percent of those who watch TV are influenced. Men get disturbed. They can't help themselves when they see Scargill, Foot and company . . . I'm becoming less and less involved with my men. They come in in the mornings with long faces. I feel like saying, 'Don't you enjoy coming in to work in the mornings?'" Another farmer says: "I find the most dangerous times are when the men are working together. Mucking about and doing sugar beet, they get talking together and start grumbling".

"Yes, things have changed," agreed a Norfolk farm manager of 8,000 acres, "You have to handle men more carefully because a good man can get a job anywhere. Before the war, men were regularly sacked and so they jumped to it. But now you have to be careful and it makes things difficult".

"Relationships have changed," a neighbouring farm manager confirmed, "especially on the owner-occupied farms. There's much less paternalism now . . . even my boss allows the workers to put Labour posters in their windows now . . ."

An eccentric example is cited in John McEwen's *Who Owns Scotland*. The largest estate in East Lothian in 1874, he points out, was Tweeddale with over 20,000 acres. Now it has only 1,600 acres. The present owner, the Marquis of Tweeddale, now lobster farming on the west coast, remarks in *Who's Who* that his task now is "striving to exist after dynamic socialism". But who, or what, asks McEwen, was responsible for the downfall of the Tweeddale estate?

But these are the farmers' and farm mangers' own perception of how things are changing in the countryside; the attitudes remain very different from those prevailing in industry. The more traditional approach was confirmed by a Suffolk farmer with 3,700 acres: "We have good personal relationships and pay the men while they are sick, lend them money and so on. We want to create a community, a secure and happy framework for their lives. We even have our own football club and so on, we have parties . . . The best way to handle workers is to give a sense of pride in the job . . ."[12]

These are well-meaning ambitions. The humdrum and impersonal, indeed anonymous, nature of most jobs in inudstry does encourage a one-dimensional, materialistic, attitude with higher wages the sole ambition and only objective. But the relationship between the landowner and his farm-worker is even more unequal. Workers must be kept in their place, and the happier they are made to feel, the more likely they are to stay there.

Farmers praise what they describe as their 'special relationship' with their workers, a relationship which Jack Boddy, the farmworkers' general secretary regards simply as a kind of exploitation — the none the better, he says, for being of the paternalistic variety. There are, in fact, more unfair dismissals, as tribunals reveal, per numbers of workers in agriculture and the food industry than any other sector with the single exception of the building trade.

"Gross incompetence. Wilful neglect. Constant lack of application. All here were patient beyond belief. I dare not summarise the accumulated damage that this youth caused on the farm. He is no loss at all to horticulture or agriculture." That is the reaction of one farmer interviewed for the study by Reading University for the Agricultural Training Board into why so many — one in three — apprentices failed to complete the training course.[13]

Yet many farmers have acknowledged that they have used the Government's Youth Opportunities Programme as a means of getting free labour for jobs which they would otherwise have had to hire either full-time farmworkers or casual labour. This is another way — the fact that so many farmworkers are eligible for Family Income Supplements is yet another — in which farmers are being subsidised by the State and taxpayers.

The attitude to organised labour was again reflected following a fatal accident in 1977 at a freezing plant at Thorganby, Yorkshire, owned by the Eastwood family which recently sold its massive egg business to the Imperial Group. The union had already warned about the safety standards there. As *The Landworker* puts it, Sir John Eastwood owns land there, houses, the village hall, even the local bus shelter; but the farmworkers' union is made most unwelcome.

In another incident, a farm manager was sacked by the Marquis of Normanby, the lord lieutenant of North Yorkshire who owns a 15,000 acre estate near Whitby, for living with another man's wife. Mr Hawking, who had been separated from his wife for over a year, was awarded compensation by an industrial tribunal. Lord Normanby had written a letter to Hawking which read: "It is in my opinion that you are no longer in a position to foster and maintain the farm's relationship with the agricultural community of the area or to promote the estate or the farm's interest". The tribunal was told that a somewhat feudal picture had been built up around the estate, which has about 80 tenants, and few people were prepared to talk about the way it was run. But one woman said she was disgusted by Hawking's dismissal. "In this day and age," she said, "I think it is outrageous that such a thing could happen".

The Duke of Westminster told *Woman's Own* in 1978: "Somebody once said to me that you should never become too friendly with an employee in case the time came when you had to sack him. It's much easier if you don't have a very close friendship". He described how he dismissed a gamekeeper two years earlier. "I sacked that man," he said, "because he was in a position of trust. It cannot be understood by people living in urban areas. There's a very close relationship on an estate between an owner and his gamekeeper".

There is a very clear distinction between friendship and the set, close (and unequal) relationship between the landowner and farm manager on the one side, and the rest of the farm's staff and local inhabitants on the other.

An arrogant, off-hand, attitude extends even to the law. In an unusual, but not untypical, case towards the end of 1977, a gamekeeper shot a poacher on the 2,000 acre Carlton Towers estate in Yorkshire owned by the Duke of Norfolk and managed by his brother, Lord Martin Fitzalan-Howard. The pheasant poacher, John Parfitt, was an out-of-work lorry driver. The gamekeeper, Roland Senior, had worked on the estate for eleven years and Fitzalan-Howard described him as the most valued employee on the farm. No criminal charges were made and at the inquest, the coroner, Miles Coverdale, said that people who trespassed did so at their own risk. Mr Senior told the police: "There is a saying — 'all the pheasants ever bred is never worth a man dead.'." But Fitzalan-Howard said that a gamekeeper's job was to protect property. "It is a hazard," he added, "especially with wild men coming out with shotguns".

The CLA, at least, was not happy with the coroner's attitude. "It's not the case," said Michael Gregory, its legal adviser, "that if a poacher comes on to someone's land armed, he is asking for it". Lord Arbuthnott puts it another way. Most shooting men pride themselves, he claims, on a liberal attitude towards the casual poacher who is simply taking 'one for the pot'.

Living in the country, for the small farmer and farmworker, can be a raw, unsophisticated but also brutal existence. They are cut off from consumerism, from cosmetic, frequently superficial, temptations and entertainment, but also the cultural attractions of the towns. Country life is a world away from the image perpetuated by romantic urban-dwellers, second home-owners, those who pay the occasional visit, and by advertising copy writers. A field of cows may look nice and peaceful, but cows catch diseases. A field of corn may look attractive on a sunny July evening, but farmers and rural labourers have to cope with sudden changes in the weather, and milk cows in early winter mornings.

In the country, the difference between wealth and poverty, though it may still be hidden behind pretty cottages and birdsong, is as striking as in towns, perhaps more so. Country people may be suspicious of townspeople, but they are often intolerant of each other; there is as much malicious gossip as there are skittles and beer. Small farmers, farmworkers, the odd-job men and char-women, the peasants of England, may be shrewd and resilient, yet is it right for them to be dominated by individuals whose wealth and manor house is theirs simply because it is inherited?

In a small village in Gloucestershire, the landlord waited for the elderly tenant to die. The tenant refused to pay his son, who had four children, more than the minimum rate for farmworkers — about £64 per week. The father died; the landlord, after more than a year was still refusing to offer the son the tenancy. The same landlord, lord of the manor of the village of Overbury, a very attractive village — fought for many months against the local authority's plan to widen a crossroads to make it safer, to take up just a few square yards of the estate. And then he conceded only after squeezing every penny of compensation from the council.

There is more to country life than glossy magazines found in dentists' waiting rooms, or published by the British Tourist Board. Beyond the sweat and the dirty fingernails, and behind the sweeping drives and manor houses, there are tales every bit as grizzly as those you can read in some childrens' books of fairy stories.

The vast acres of the country should be shared by the people Not attacked by envious politicians, urban crowds or bureaucrats, but opened up with an attitude of tolerance, social justice, equality of opportunity, imagination and enlightenment — in short, those qualities so rare in rural areas today.

Travel, modern communications, capitalist farming, have meant that people no longer help each other; they no longer need to rely on each other; they do not know each other. All sense of community is vanishing. There should be no illusions about the past, or notions about 'Merrie England' when the gap between rich and poor was even starker. But now is the time to marry the best of the new attitudes with the best of the old. If we put our minds to it, we can both master technology, begin to respect Nature again and think about the importance of community and the land.

11

THIRD WORLD PERSPECTIVES*

"Oil is the foundation of many of the components of our present civilisation; petrochemicals, plastics, synthetics and the like. Therefore, in a sense, our generation is busily burning away part of the basis of its own civilisation." – *Michael Manley, former Prime Minister of Jamaica.*

"Washington would acquire virtual life and death power over the fate of the multitudes of the needy . . . not only the poor, less developed, countries, but also the major powers would be at least partially dependent on food imports from the United States." – *The CIA in 1974, a year after the so-called world food crisis.*

The increasing demand for food and a totally unjust food trade system are, two of the most serious threats facing us. Another is the Middle East. They share similarities. One super power — the US — dominates world trade in food; a small group of countries dominates trade in oil. A relatively small group of large multinational corporations control trade in both food and oil.

The Middle East is constantly in the headlines and the concern of governments and their diplomats. The iniquities and dangers inherent in our present food system are rarely discussed. Few people are starving in the West. We have other preoccupations. Yet the control enjoyed by the US and large

* by the Third World, I refer to non-oil-producing developing countries and, in the context of this chapter, the people, and especially the poor, in these countries, as opposed to the governing elites.

companies is concentrated as much in the area of food supplies as in the supply of oil. And there is another link between these two basic commodities. The determination of large companies, in the interests of short-term profits, to impose an agricultural system, and even a diet, on the rest of the world, also ensures that oil-based technology, despite the obvious costs and dangers, increases its hold over food production in more and more countries.

Just as self-reliance is a goal we could all seek in the interests of social justice and peace, so it is even more important for a Third World becoming increasingly dependent on the West for its basic needs and exploited by a sophisticated capitalist, trading system. And the inequalities in Britain of wealth and land ownership are reflected, though on a much more dramatic and significant scale, in developing countries.

A mere 2.5 percent of landowners with holdings of more than 250 acres control nearly three-quarters of all the land in the world, and the top 0.23 percent control more than half. Yet about 70 percent of the total population of the Third World make their living from the land. In Zimbabwe, to take an example of interest to Britain, nearly half of the country's land, including the best, was at the time of independence in the hands of about 6,000 white farmers; the rest was shared by 3 million black peasants.

It is a startling fact, given the number of people who are either hungry or starving, that since the Second World War, food production has increased faster than the growth in population (though the rate of increase in food production is now slowing down). Between 1949 and 1975, food supplies increased by 155 percent; the population rose by 67 percent. Over the same period, the amount of rice available per person doubled. During the ten years to 1977, food consumption *on average* grew much faster in the developing world than in the industrialised world in terms of both calories and protein. The UN Food and Agriculture Organisation acknowledges that the increase in poverty has been associated not with a fall, but with a per capita rise, in cereal production.

Why, with consistent bumper harvests, is there scarcity and poverty in the midst of plenty, with the rich getting richer and the poor, poorer? Largely because of international agricultural systems, whether they go under the guise of 'free trade' — a euphemism, as far as grain is concerned, for domination by the US — or whether they are protectionist.

Food is a weapon. While President Carter criticised America's record as an exporter of arms, he had no qualms about America's role as the world's largest food producer. When President Reagan was sharply criticising Moscow, he approved a record US grain sales deal with the USSR. It is a weapon used more visciously than ever before. It is wielded by the US whose balance of payments would be in persistent deficit without its farm trade, and which accounts for well over half of all world grain exports. It is deployed just like any other commodity in the interests of the rich in an open, capitalist economic system in which the Soviet Union, the leading Communist power, is also an active and energetic participant. Hunger is thus part of the power game, the result of poverty land politics rather than the shortage of food in the world. Warnings of imminent food crises and of 'over population' are echoed enthusiastically by vested interests, notably farmers' organisations and agrochemical companies determined to defend their right to expand their profits and their share of the tempting world market. The power of the American agricultural lobbies, coupled with the cost of guaranteeing farmers' prices when there is a slump on the world market, ensures that a grain embargo against the Soviet Union is in neither the political, nor economic, interests of Washington. So whatever the rhetoric of Washington and Moscow and, indeed, the reality of the arms race, food will remain a feature of East-West relations, especially at a time when the Soviet leaders are determined to try and satisfy their consumers' increasing expectations. The immediate needs of the Third World, will therefore be accorded very low priority.

As more land is cultivated and yields increase, developing countries have become more, not less, dependent on food imports from the rich industrialised world. Years ago, Egypt and Bengal were productive, rich, countries, the granaries of their regions. Now Egypt imports rice and grain from the US (America has become the world's largest rice exporter) and sells new potatoes to Britain. In the 1930s, most areas of the world, excluding Europe, were self-sufficient in grain. In

Changing patterns in the grain trade*

Million metric tons

Region	1934-38	1948-52	1960	1966	1973**
North America	+ 5	+23	+39	+59	+91
Latin America	+ 9	+ 1	0	+ 5	− 3
Western Europe	−24	−22	−25	−27	−19
Eastern Europe and USSR	+ 5	−	0	− 4	−27
Africa	+ 1	0	− 2	− 7	− 5
Asia	+ 2	− 6	−17	−34	−43
Australia and New Zealand	+ 3	+ 3	+ 6	+ 8	+ 6

* Plus sign indicates net exports, minus sign net imports
** Estimated

Source:
US Department of Agriculture

1960, Zambia exported food; it now spends a third of its export earnings from copper importing food. Sri Lanka was self-sufficient in food until it was used as a supplier for Britain's tea drinkers. (It again became self-sufficient faced with the emergency of the Second World War.)

Colonial powers ruthlessly used developing countries as suppliers of raw materials, of both minerals and commodities such as coffee, sugar, cocoa, jute and cotton which could then be processed by industrial workers and consumed cheaply in Western Europe. They promoted a system of monoculture and a plantation economy that directly contributed to the poverty, dependence and hunger of the local population.

When the political empires crumbled, the commercial and economic domination of the West, now joined with a vengeance by the US, perpetuated the system so that the Third World is encouraged and persuaded to concentrate on exporting cash crops to the West at the expense of providing food to meet the needs of the local inhabitants.

Africa is a net exporter, not only of palm oil and peanuts, but also of barley, beans, cattle and vegetables, yet suffers worse malnutrition than any other continent. Mexico exports cattle and asparagus to the US, yet 80 percent of the children in rural areas are under-nourished. The American-based House of Bud has set up plantations in Senegal to supply Europe with winter vegetables. Up to 90 percent of Peru's anchovy catch is exported to the West and Eastern Europe as fishmeal to provide feed for intensive chicken farms. During the severe drought in Ethiopia in 1973, businessmen on a UN-sponsored trip bought up local cattle for Europe's pet food industry. The United States, the world's largest exporter of staple food, is buying more food from the developing countries than it is from other industrialised nations while the Third World in turn is buying more and more basic food from the US. The developing countries' food import bill increased by £3,000 millions between 1973 and 1974 after the Soviet Union had 'raided' the American grain silos. Their grain imports rose from 20 million tonnes in 1960 to 50 millions in 1970 to 80 millions in 1978, and was expected to reach over 100 millions in 1982.

There is nothing wrong in principle with international trade; after all, it can promote peace, mutual interests between countries and even a better quality of life for ordinary people. But there is a great deal wrong with a trading system dictated by the short-term financial and political interests of the richest countries and corporations. The price of food is set by the American government and grain companies (with the help, from time to time, of Moscow). Washington has to react every year to pressure from American farmers for high prices as well as to political and diplomatic considerations about whether or not it should export grain to feed cattle and poultry in the Soviet Union. In an attempt to push up the world market price of grain, and preventing it from subsidising farmers from its own funds, Washington in 1972, just a year before the food crisis, paid its farmers $60,000 millions not to use sixty million acres, more than 15 percent of US cropland. Prices rocketed and following poor harvests in the Soviet Union and Europe the next year, the Nixon administration imposed an export embargo on soyabeans, a commodity Washington normally insists should enter the world markets with no restrictions at all.

In 1977, after two years of bumper harvests, Washington ordered American farmers to cut their wheat acreage by 20 percent; it was planning to do so again in 1981/82. At that time, stocks of wheat in the US stood at about thirty-eight million tonnes, significantly more than the amount required, according to both US and UN calculations, to alleviate the hunger of 500 million people in the world considered to be suffering from 'absolute hunger'.

In 1981, faced with the prospect of another record harvest and further subsidies to maintain farmers' incomes, Washington indulged in a massive campaign trying to persuade the world, and developing countries in particular, to eat more wheat. The campaign, which included setting up model bakeries in China and giving away loaves of bread in the streets of Tokyo, was organised by US Wheat Associates, an organisation financed by Washington, by US wheat farmers and by the governments of 92 countries in which the promotion is carried out. The idea was to show these countries how to make biscuits, bread, and even doughnuts (the promoters earned the name 'doughboys') in such countries as the Philippines and Taiwan. Meanwhile, the National Biscuit Company of America has launched a campaign to sell Ritz crackers in the Third World. It has also developed Ricetein, a rice-soya mixture, with the company explaining: "It is priced higher than rice, but is less expensive than meat".

But the US is imposing a diet of meat as well as wheat on developing countries, ironically just at a time when meat consumption for both health and financial reasons is declining in the industrialised world. The reason? — to expand the market for American feedgrains. The consumption of chickens, in particular, is being promoted. Of the 1,250 million or so tonnes of food and feed grain produced annually in recent years, developed countries eat half, though they account for only a quarter of the world's population, and their animals eat a quarter of the grain, equivalent of the total human consumption of China and India put together. And of the grain consumed in the West, seventy percent is used for animal feed; the Third World traditionally fed only ten percent of its share of the world's cereals to animals.

But the picture is now changing: more grain is now fed to animals than is consumed directly by people in poor, developing, countries. (In richer, developing, countries such as Singapore or Mexico, the proportion fed to animals is much higher.) Over the past decade, the growth of grain consumed as animal feed has been twice as great as grain consumed directly by people, and since 1960 the percentage of grain for livestock has doubled, from 20 percent of total world production to 40 percent.

Imperialistic presumption, the assumption that what was good for their own population is good for others', or even a naive belief that donors of aid were doing good, encouraged cattle ranching in such areas as the Sahel in West Africa, the Kalahari desert and Ethiopia, thereby depriving land for more basic food and compounding the problems in times of drought. In Kenya, the World Bank regards the pastoral tribes as future meat producers for European markets and encourages subsistence herders like the Masai, Samburu and Turkana to become commercial ranchers.[1]

The West is imposing an agricultural system on the rest of the world to suit its own interests. It regards the Third World as a vast untapped market, of both animals and humans, for its surpluses of grain. If an American-style diet, based on intensive meat production, was progressively imposed on the rest of the world's inhabitants, available cultivated land would soon be exhausted.

The rich countries use the poor as providers of food or of raw materials — hence the traditional imposition of monoculture and large plantations in the Third World (only to be rejected and dismissed when a cheaper substitute is discovered). At the same time the rich control food prices through their domination of world trade.

One of the problems is that the cost of Western-style

WHOSE LAND IS IT ANYWAY?

farming is high, and is increasing, because it is energy-intensive. So those farmers will demand prices which the poor in developing countries will not be able to afford. Meanwhile, aid programmes further exacerbate the problems of local farmers in the Third World by keeping down the price of food there, so depressing the income of local food producers. That problem is compounded by the drift away from rural areas to shanty towns near the cities. These are already full of unemployed people of whom local rulers are frightened and, as a result, try to appease by subsidised food prices. In addition, the terms of trade — that is to say, the relative prices of exports and imports — persistently work against the interests of raw material or food exporters in developing countries so that they have to export more and more food to pay for increasingly expensive manufactured goods, such as tractors, produced by the industrialised world. In short, as the Economist Intelligence Unit concluded in its report on the world grain market in 1980: "in essence, only the Third World can feed itself".

Sheltered behind governments, in both the developing and developed world, lie the large, unaccountable, international food corporations. The five giant international grain traders — Continental, Cargill, Andre, Louis Dreyfus and Bunge — showed at the time of the so-called 'great grain robbery' in 1973, and again three years later, how they deal with foreign governments, in this case, the Soviet Union, without Washington knowing what is happening or being able to do anything about it.[2]

Del Monte, the large US-based fruit and canning company, owns farms and factories in more than twenty countries. For instance, it grows asparagus in Mexico for American consumers. Other large corporations prefer not to take the risks involved in farming themselves. We have already noted that Nestlé has become the world's second largest food company without owning a single cow or acre of coffee or cocoa plantation. Unilever, the world's biggest importer and user of edible oils and fats, is getting rid of property and plantations in many countries, preferring the flexibility and freedom to play the commodity markets without being tied down to any particular source of supply. (Though the company is still holding on to plantations in 'safer' countries, its subsidiaries dominate the economy of the West African state of The Gambia.)

Large companies are operating side by side with the rural poor; indeed, they exploit both the low wage structure and the political and social inequities perpetuated by Third World governments as well as these governments' hopelessly weak bargaining position. Just 10 percent of the price of bananas in greengrocers in the West finds its way back to those who watered and harvested the product in the first place. Of every £ we spend on instant coffee, nearly half never leaves Britain and, of the rest, the bulk goes to estate owners, middle men and shippers. The twenty million or so people who tend the coffee in the plantations each receive on average less than £45 a year.

A handful of British companies control the trading, blending and packaging of tea. For this, they get 25 percent of the retail price. Investigations by the World Development Movement show that in the British-owned Tata-Finlay estates in Assam, North-east India, infant mortality almost doubled between 1973 and 1979. In Kenya, where Finlay and Brooke Bond own large estates and where tea yields doubled over a three-year period, wages earned by tea workers were barely enough to feed a family of five. In Malawi, tea workers on plantations owned by Ruo Estate Holdings, a British-based company, are paid 15p a day, the price of a loaf of bread. After a 90 percent rise in Brooke Bond's profits and a 32 percent increase in dividend payments to shareholders, basic wages in the North Indian estates were increased by just 10 percent in early 1978.

Now large multinational corporations are tightening their grip on agriculture's most basic requirement — seeds, the most fundamental resource. Plant breeders' rights, guaranteeing patents through legislation, protect companies which have developed hybrid varieties, and encourage them to take over as much as they can of the lucrative market in seeds. At the same time, they promote uniformity and threaten the genetic diversity which has proved so vital not least as a source for drugs to help counter disease.

The system works like this: companies develop a hybrid plant variety with certain characteristics, a high yield for example. That variety is then patented either with the national seeds authority or with the EEC. The governments and the EEC, through its common catalogue, accept only patented varieties, thereby banning the cultivation and marketing of non-patented varieties. It is easier for companies, and easier to check patents, if varieties have standard, uniform characteristics.

One advantage to the companies is that hybrids must be purchased every season; in many cases they cannot produce their own seed stock. Thus farmers in both the West and in the Third World become dependent on the large seed companies, to whom they have to pay royalties every year — in addition to the basic cost of the seed — and are tied to companies through long-term contracts. It is no accident that many of the major seed companies are also chemical and drug producers. Hybrid seeds, especially the high-yielding varieties, need expensive applications of pesticides and fertiliser, which the companies also produce. Their interest in seeds also complements their role as drug companies. They have it all ways; the perfect, all-round, commercial operation.

We have noted that as soon as it discovered that patent legislation was about to be introduced in the 1970s, large chemical firms including Ciba-Geigy, Sandoz, Pfizer, Monsanto and Union Carbide entered the seed market. Shell is now the largest seed distributor in the world. Of the 562 patents issued by the US Plant Variety Protection Office up to March 1979, nearly half were granted to the largest 17 seed company buyers, less than 10 percent to universities or independent agricultural research stations. Two-thirds of all hybrid maize varieties in the US are supplied by only four companies.

The international seed market is worth more than £5,000 millions a year. By 1980, poor countries were paying out millions of pounds to Western-based companies for seeds, many of which were derived from natural, indigenous, or wild genetic materials collected from the Third World countries themselves; for the origins of virtually all basic crops lie not in the US or Europe, but in developing countries and the Mediterranean. Wheat came from Ethiopia and Asia Minor; rice from South-east Asia and West Africa; maize from Central and South America; broad beans, carrots and peas from Central Asia; sugar cane from South-east Asia, and apples from Central Asia.

American companies are selling hybrid varieties of rice to Asia and Africa. As Dr Garrisson Wilkes, specialist in corn genetics at the University of Massachusetts, puts it: "Suddenly in the 1970s, we are discovering Mexican farmers planting hybrid corn seed from a Midwestern seed firm, Tibetan farmers planting barley from a Scandinavian plant breeding station, and Turkish farmers planting wheat from the Mexican wheat programme. Each of these classic areas of crop-specific genetic diversity is rapidly becoming an area of seed uniformity".[3]

There are about 300,000 higher-order plants: historically — and, indeed, prehistorically — some 3,000 species of plants were cultivated for food. Now only 20-30 species provide the entire world with staple food: such crops as rice, wheat, cassava, soyabean, barley, bananas, sugar cane, potatoes and peanuts provide ninety percent of all human energy. Of these, just three, wheat, rice and maize, supply the bulk of the world's seed crop.

But the process, dangerous but scarcely noticed, has not stopped there. As entire species are threatened, so are the different varieties of others. For example, over the past decade, more than 100 varieties of Brussels sprouts alone have vanished from breeders' shelves and catalogues. Some fifty years ago, eighty percent of the wheat grown in Greece consisted of native breeds; now, more than ninety percent of the old, traditional, strains have virtually disappeared. Thousands of varieties were grown by small farmers in the mountain areas of Greece. New hybrid seeds were developed that could grow in the valleys. The moutain farmers were made redundant, and emigrated to the cities. Not so long ago, farmers in Turkey sowed their fields with a variety of different seeds held over from the previous harvest. They now buy uniform varieties every year from large seed companies. Plant gene pools in traditional centres of crop diversity, including Turkey, the Middle East, Afghanistan and India are shrinking dramatically.

These companies, and research institutions in both the

West and East, are busy scouring developing countries for traditional varieties with disease resistance strains from which they can breed their hybrids of the future. Varieties are vanishing fast, and there is increasing concern among some scientists that valuable varieties will vanish forever. The UN Food and Agriculture Organisation itself acknowledges that the European Common Market's common seed catalogue threatened to wipe out three-quarters of Europe's vegetable varieties by 1981.

British plant breeders were recently looking for resistance against a new form for lettuce mildew. No cultivated varieties were resistant to the disease. Eventually, one was found growing wild in Turkey. It is a familiar story.

New hybrid strains, with their genetic uniformity, are more vulnerable to disease. In 1970 the southern corn leaf blight wiped out 50 percent of the maize crop along the American Gulf Coast, a disaster which pushed up prices on the world grain markets. At about the same time, a little noticed, but more catastrophic, mildew epidemic nearly halved the pearl millet crop in India. The cause was the same: the planting of the same variety of seed over millions of acres. When a mould hit twenty percent of Zambia's maize crop in 1974, traditional, native, varieties were virtually unscathed. The potato disease which starved 1 million people in Ireland in the 1840s had already demonstrated the dangers of relying on one crop, the dangers of monoculture.

It is a viscious circle: as breeders develop a new hybrid, it is attacked by a pathogen. A resistant variety is added to the original. That, too, soon becomes vulnerable to attack. "In their wilderness state," Wilkes points out[4], "both plants and diseases which attack them are forever adapting to each other through the evolutionary process. The diseases mutate new forms of attack, the plants new forms of resistance . . . Under modern agriculture, plants no longer mutate but are grown from new seeds each year for continuous high yield. The mutation of diseases, however, cannot be stopped".

A similar message was delivered at the Royal Society of Arts in 1979 by two of Britain's more thoughtful plant scientists, Martin Wolfe, head of the department of pathology and entomology at the Plant Breeding Institute at Trumpington, near Cambridge, and J.A. Barrett, a researcher at Cambridge University's department of genetics. "Current methods of agricultural production", they said, "offer a perfect ecological niche for disease and pest organisms; it is vast, rich in nutrient, and is continuously available, so that particular forms of pathogenic fungi, for example, which possess the ability to grow a particular host variety, can do so without restriction, over perhaps millions of acres".

Aware of the dangers of genetic uniformity, scientists are attempting, with genetic manipulation, to secure the best of both worlds: develop seeds which combine the modern, commercial, demands for high yield, and disease-resistance. But Wolfe and Barrett noted that "considerably more diverse mixtures have been widely used in subsistence agriculture for thousands of years", and that, even today, traditional varieties of grain even from the point of view of yield, are more effective, and even more profitable for bread-making and for malting barley.

What is happening to trade in seeds — the growth of a kind of 'genetic imperialism' — has passed virtually unnoticed, even by farmers themselves, it epitomises so much what is wrong with modern agriculture and illustrates the power large corporations and landowners now enjoy.

Concern about the increasingly dangerous trends and the disposal by seed merchants of their stocks of traditional varieties led to the establishment of the world's first important vegetable gene bank by Oxfam at the National Vegetable Research Station in Wellesbourne, Warwickshire, in October 1980. Brian Walker, Oxfam's director-general, said at the opening: "In recent years, the failure of seed companies effectively to market seeds at prices which the poor can afford, has been a constant refrain in reports from our field staff. These difficulties were compounded by the disappearance, increasingly, of natural land races, and of species commonly used by ordinary people, as pollution and the so-called green revolution took its toll".

Pointing to the political importance of sharing genetic resources, he added: "Ultimately, food is a more powerful weapon than the bomb, and within the food syndrome, whoever controls genetic resources can wield immense political power".

The arrogant exploitation of the world's natural resources extends to forests. Forests, and tropical rain forests in particular, now contain about half of all the wild species left in the world. They are a highly complex and delicate ecosystem, providing and creating a home for an immense variety of species which are also a source of medicines, including anti-cancer drugs. Scientists are still collecting hundreds of new plant species from Africa. A single volcano in the Philippines contains, on its slopes, a greater variety of woody plant species than the entire United States. Less than half of the species have yet been catalogued. The Amazon basin contains about 1 million different plants, making it the richest region, biologically, on earth. Yet it is being destroyed by bulldozers. An area of forest the size of Wales is cut down every month. The Amazon jungle is being placed, by the Brazilian government, at the mercy of Western-based consortia, including Vatican funds, anxious to cultivate profitable cash crops and set up vast cattle ranches. Yet it is estimated[5] that the Amazon forest contributes, through photosynthesis, 50 per cent of the world's entire annual production of oxygen. And it is well known that all forests and trees absorb carbon dioxide.

Tree-felling has already begun to influence the climate. In Bolivia, winds are increasing in strength and there is less rainfall; in Peru, there is less snow on the Andes. In the forests' extremely sensitive ecosystem, trees live almost independently from the soil — the layer of humus is very thin — and trees recycle half the rain that falls on them. In some areas the temperature has increased by 30° F, and rainfall reduced by 5 per cent, as a result of cutting down large areas of forest. It is widely assumed that a rise in the amount of carbon dioxide has led to dryer weather in the Northern Hemisphere, including the American grain belts. Meanwhile, the Government of Indonesia, with the second biggest area of tropical rain forest, has leased the bulk of the land to international timber corporations. South-east Asian rain forests currently supply 85 per cent of the world's demand for timber.

We know more about the moon than about some of the rainforests on earth. Yet by 1976, more than forty per cent of the world's original tropical forests had been destroyed. At the present rate of destruction there will be little rainforest left in fifty years, just the time when many oil deposits are expected to run out. While deforestation is leading to climatic changes, including floods in some regions, deserts are creeping outwards in Africa, North and South America, India and Indonesia. At least 50 million people live in areas where inappropriate and modern farming methods, including intensive livestock production, are destroying the land. In the Sudan, the desert has moved southwards 62 miles in less than two decades; in North Africa, more than 250,000 acres of farmland are lost to the desert each year; and according to an official survey, more than 50 million acres of land in the US are in poor condition because of overgrazing while a third of all American cropland is suffering from severe soil loss.

In the Third World, the problem is aggravated by the continual search for firewood. It is compounded simply by lack of thought. For example, if animal numbers in West Africa were halved, with the use of slightly more sophisticated husbandry there is little doubt that meat production could be doubled. But the problem is essentially political in the broadest possible meaning of the word, that is to say, the result of policy. Controlled grazing, for instance, and irrigation techniques in Israel, Algeria, China and Somalia have demonstrated that it is both financially and technically feasible to stop this deterioration of the soil.

One major scientific development, initiated by, and the product of, research by those with the best will in the world, has contributed to these dangers, — to the threat of genetic diversity, to the reality of increased control by large landowners at the expense of the poor, and to the concentration of landownership. That is the so-called green revolution. It was based on the breeding of hybrid, high-yielding, dwarf varieties based on a single parent plant. It led to a significant increase in world food production, enabled India to feed herself, raised rice production in many other Third World countries and opened up new lands for cultivation.

But these varieties are particularly vulnerable to disease. They require large amounts of fertiliser and pesticides and expensive irrigation equipment. The cost to India of being able to feed itself (in most years) in this way is an import bill of £2,500 millions' worth of capital equipment and agrochemical materials. Rich farmers have thus been able to benefit, but not the poor majority. Unemployment rises sharply.

In Sonora, Mexico, where the green revolution started, landless peasants used to account for 57 percent of the workforce; now they account for 75 percent. The Asian Development Bank, which enthusiastically promoted the green revolution, now acknowledges that rural unemployment has increased, while the number of people who are under-nourished, in spite of the increase in food production, has also risen. "The region", the Bank reported in 1977, "is no closer to solving the food problem than ten years ago". The story is the same today. Landless peasants now account for more than half of the rural population in Third World countries.

The green revolution has stimulated capital-intensive agri-culture in developing countries, a development that is singularly inappropriate, given the shortage of available capital and the abundance of labour. In India it has led to the eviction of tenants and sharecroppers. There are echoes of the Enclosures in Britain. Yet it need not have been done this way. South Korea, for example, has shown how high-yielding crops can be adapted and included in comprehensive land reform. Even the World Bank now recognises that land reform is crucial.

China, Taiwan, South Korea — the latter by expropiating estates left by Japanese settlers — have demonstrated that a combination of land reform, elements of the green revolution, good husbandry and labour-intensive farming methods can produce the best yields in the world. The small farmer has almost everywhere proved to be the most productive. To take just one example: in Colombia, though small farmers work only a quarter of the cropland, they produce two-thirds of the total agricultural output. If they combine to form cooperatives, so much the better.

Other countries have followed a different path to those in the East. Partly because of the green revolution, there has been an increase in food production well above the rise in pop-ulation. Yet per capita food consumption has barely increased at all over the past decade in developing countries. During the past ten years, Brazil recorded a massive increase in food production; but per capita protein consumption actually fell 5 percent. Brazil has concentrated on growing cash crops, including soyabeans for export and animal feed, and sugar cane for fuel. Thailand over the same period has increased its tapioca (cassava) crop from 3 million tonnes to 13 million tonnes, yet the crop is used mainly for animal feed and has been bought in such large quantities by European livestock farmers as a cheap substitute for high-priced European grain that EEC cereal producers have complained; much as European car-producers complain about the flood of Japanese car imports. Meanwhile, nutritional standards in Thailand in terms of calories, protein and fats, are falling.

Breakthroughs are coming in biotechnology and genetic engineering, the production of bacteria-based protein and the artificial manipulation of genes. For the former, Western-based companies have already said that they regard the Third World as an ideal test market. As far as genetic technology is concerned, developments could indeed lead, for instance, to the creation of plants with nitrogen-fixing properties that would obviate the need for fertilisers, or of plants with a higher rate of photosynthesis. But to be able to benefit, developing countries will have to tie their hands even more firmly to the rich.

If there was a true, fair and agreed international division of labour, then international trade could promote the interests of all. But today this seems like a utopian dream. The twin motivations and priorities, as far as the industrialised world is concerned, are consumption and profit. And their interests are allied to those of the governing elites in Third World countries, which prefer to seek refuge in aid and prestigious military programmes and one-sided trade deals, rather than helping the majority of their inhabitants. The industrialised world imposes its own standards on developing nations and at the same time secured for itself the lion's share of raw materials, many of which are, by right, Third World resources.

It is not as if the increase in the number of landless people in developing countries has led to the growth of industrial employment and generated the spread of urban wealth. Far from it; it has led to an explosion of shanty towns and underemployment shifting away from rural areas to a squalid, urban environment. It is reminiscent of Britain 150 years ago.

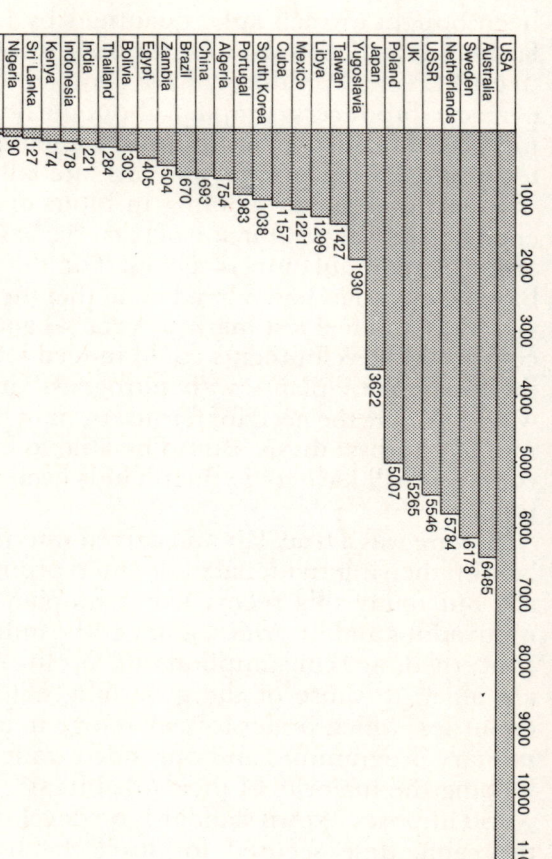

25. **Energy Consumption**: *The figures are for 1975, in kg of coal equivalent per head of population. If every field in the world was as intensively cultivated as the average field in Britain, about 50% of the present world fuel consumption would have to be devoted to the production of food. Western agricultural chemical companies are now pouring their products into the Third World, including those banned in their own countries. (Source: World Bank)*

Country	Value
USA	10999
Australia	6485
Sweden	6178
Netherlands	5784
USSR	5546
UK	5265
Poland	5007
Japan	3622
Yugoslavia	1930
Taiwan	1427
Libya	1299
Mexico	1221
Cuba	1157
South Korea	1038
Portugal	963
Algeria	754
China	693
Brazil	670
Zambia	504
Egypt	405
Bolivia	303
Thailand	284
India	221
Indonesia	178
Kenya	174
Sri Lanka	127
Nigeria	90
Haiti	30
Ethiopia	29

Multinational corporations, encouraged by Third World bureaucracies eager for investment and export promotion, continue to exploit cheap labour in developing countries.

The Earth's total resources will not be able to accommodate the spread of the West's high-energy consumption to 'underdeveloped' countries. The rich — and this includes the Soviet Union — give more grain to livestock than the amount consumed by the human and animal population of the Third World in its entirety. The US uses more fertiliser on tennis courts and lawns than India, with three times the American population, uses for all purposes. 90 percent of the protein concentrates used by British farmers to feed their livestock is imported from developing countries. Britain, with just 1.5 percent of the world's population, consumes 10 percent of the world's supply of fishmeal. Dogs and cats in the US eat more than £1,000 millions worth of processed food annually. The West's consumption of oil and other minerals is out of all proportion to its percentage of the world's population. If every field in the world is intensely cultivated up to the British average, about fifty percent of present world fuel consumption would have to be devoted to food production.

The rich want to keep their cake and eat other people's too. They have scoured other people's countries; the British Phosphate Commission, for instance, told the people of Ocean Island, the Banabans, to go off to Fiji so that it could mine the island naked in 79 years. Western commercial interests will continue to scour their own wildernesses: Scotland, Alaska, Australia and Greenland.

They have treated human resources no better than they have the land and what it has provided. It is little different today from yesterday when the rural population of Wales was uprooted to work in the coal mines of the South, only to be dropped and forgotten barely a century later.

Even the sober Economist Intelligence Unit concluded in a report on commodities in 1979: "In present circumstances, in the midst of a technical revolution in agriculture and with growing presure from Western food industries on the land resources of the developing world, it seems that you cannot both have a free market in agricultural commodities and a free market in agricultural land, because the social consequences of their combination are too terrible".

There is a mass of alternatives — the creation of village woodlets in developing countries, much greater use of biomass for fuel, more careful, appropriate and less wasteful irrigation systems, mixed-integrated-farming, rural industry, land reform. But nothing will happen unless enough people in every country fight those who control large concentrations of capital, and through that, the market.

Just at a time when energy costs were at a premium, ICI began developing a single cell protein from natural gas for animal feed, while BP experimented with the use of oil-based bacteria to feed salmon on fish farms owned by Unilever. Petro-chemical companies are cultivating yeast, fungi and bacteria on expensive energy-based products for feed in the attractive markets of the Third World which does not want it or need it.

Some interesting developments are taking place. Fifty years ago, Henry Ford said: "There is enough alcohol in one year's yield of an acre of potatoes to drive the machinery necessary to cultivate the field for 100 years". Ethanol, which is derived from high starch crops and carbohydrates such as potatoes, sugar and wheat, is being produced from plant waste. Volkswagen and Chrysler are working on it; Fiat is developing its Totem engine which can use 80 percent raw alcohol as fuel. The French National Institute of Agricultural Research claims that by the year 2025, biomass — a mixture of vegetable residues, straw, maize kernels and sometimes pine needles are added as well — could provide 25 percent of France's energy needs.

Brazil is already using alcohol made from sugar cane and manioc for 20 percent of its fuel without making any significant changes in engines. It plans to use nothing but alcohol fuels by 1985. Tractors fuelled by vegetable oil have been successfully used on farms in Zambia. Brazil estimates that less than 2 percent of its land area could produce enough fuel to replace imported oil. But what is the real cost? The middle-class of Sao Paulo "is driving to work on the product of land that could be growing subsistence crops. Petrol tanks are full, bellies empty".[6]

Food for fuel is enough of an irony. But an essential raw material for the contraceptive pill, diosgenin, also comes from a plant. Wild yams growing in Mexican rain forests are a main source of the product. Private industry bulldozed in; the Mexican Government then decided to control the production of the yams. Prices rocketed and it was also soon discovered that the yam, a delicate fruit, grew well only in its natural, extensive, ecosystem. Now an Englishman, Dr Roland Hardman of the School of Pharmacy and Pharmacology at the University of Bath has discovered that there is another source of diosgenin, the herb, fenugreek.

Yet another irony: fenugreek is a herb whose medicinal properties — and it was also claimed to be an aphrodisiac — were recognised a long time ago. It is grown mainly in the Middle East, India and China. Fenugreek, which is rich in protein, also has potential as animal feed.

12
NOW OR NEVER

"The true conflict of our time is between the manufactured and the organic world, the drive for the 'conquest of nature' or the conquest of the machine. "– *Dora Russell in "The Tamarisk Tree".*

The world has enough for everyone's needs, but not for everyone's greed." – *Mahatma Gandhi.*

We are facing a gigantic political, economic and social crisis. Representative parliamentary democracy is under threat: MPs have no power and little influence when faced with the massed ranks of local and central bureaucracy and powerful interest groups, including the large multi-national corporations. Our industries are crumbling, apparently irrelevant to new economic and commercial realities. Investment is falling; unemployment continues to rise. A kind of capitalistic anarchy is destroying what is left of our communities, neighbourhoods, even pride in the country and in our towns. For we are also facing a crisis in values: successive governments have encouraged an attitude of "beggar thy neighbour", of allowing the wealthy to continue to enjoy their privileges and powerful groups, including trade unionists and City bankers, to wield the strength they have, even if that means trampling on the interests, prospects, and the liberties of those less fortunate, weaker, or less organised around them. They have encouraged a frightening degree of materialism, of greed and consumerism. We are living in an acquisitive society. Not only in terms of disposable income but also in terms of the environment and the ability to enjoy leisure is the gap between the better-off and the poor widening. This is strikingly evident in the country where inherited manor houses, second homes and even first homes sit side-by-side with grinding rural deprivation, and where manicured lawns overlook poisoned pastures, toxic waste tips and polluting motorways.

Britain is becoming increasingly and frighteningly polarised between those who can survive easily on money and power and those who feel more and more alienated from the criteria and standards of a late industrial, capitalist society. There are many directions this polarisation can take. Europe 2000, an EEC-sponsored project which cost £1 millions and paid 200 experts, forsees a greater demand for peace as well as more terrorism and urban violence; more corporatism at the national and international level, but more local autonomy; more highly concentrated areas of industry and population, and more dispersal; a more educated public and more resistance to change from groups, professions and organisations with power. "Above all," the study concludes, "society will be more puritanical."

Another report, Europe 2002, Britain plus 25, drawn up by the Henley Centre for Forecasting prophesies growing conflict between meritocracy and populism, more money spent on law and order to quell the tensions; some kind of national service providing a large class of people with state employment — more than an echo of Orwell's *1984*. That study concludes: "Hunger, like wars, unfortunately appears to be a permanent feature on this planet. Eradication of this problem would require a substantial sacrifice of living standards in the richer countries, and this is most unlikely". Meanwhile, the French Association Internationales Futuribles bluntly observes: "The social melting-pot will in effect be replaced by a process of polarisation separating those enjoying the benefit of progress from those left behind by expansion".

One of the main scandals to which we have tried to alert the reader is the deep and continuing inequalities in our society and the great concentration of wealth. Another is the lack of public debate about long-term problems, and the blatant inconsistencies, dishonesty and wastefulness of those with the power and authority to make decisions which affect the rest of us. The inconsistencies are an insult to common intelligence and would amaze a visitor from another planet. Governments

— especially the Thatcher administration — insist that they want to reduce the role of Government, to make it more accountable, to decentralise democracy and the administrative machine. In fact they impose more restrictions on local authorities and tighten the grip of the centralist state in such areas as "law and order". Computers have tags on all of us.

Office blocks stand empty, yet the equivalent of 100 "Centre Points" in London are in the process of being constructed. In spite of this and the number of homeless and those living in tower blocks they desperately want to leave, unemployment among construction workers is running at about 25 percent, almost double the rate in any other industry. Unemployment pushes up public expenditure (3 million unemployed costs the Government about £200 million a week in benefits and lost tax revenue) and involves greater bureaucracy, though the Government is determined to cut public expenditure and the number of bureaucrats. Britain is sadly in need of investment, yet £4.5 millions worth of capital was exported within less than two years of the Government's decision to abolish exchange controls. The development of new transport systems and new technology should lead to the country becoming a smaller place, to easier access to more areas; in fact, traffic jams increase, decisions are taken even further away from where we live. We are becoming more and more alienated, while the British establishment remains as insular as ever.

Inconsistencies, totally irrational behaviour, are the result of a failure to think, to look back, to look around, to survey the scene. They reflect short-term interests and the concerns of ambitious party politicians, senior civil servants and financial managers, as well as the failure of banks, with their enormous profits, to invest in "risky" initiatives and imaginative ventures.

This unaccountable hierarchy spends millions of pounds of public money in prestigious projects such as Concorde, employ thousands of people in tank production only to lay them off when the buyer (in this case, the Shah of Iran) is deprived of authority and power. The lack of any coherent long-term policy is evident, in the way labour in agriculture has been replaced by energy-intensive and capital-intensive investment.

Just a few days after Dupont, the giant American textile and chemical group, announced with great applause a decision to invest £29 millions in a new plant in Northern Ireland — a move that was acknowledged would lead to a permanent job loss when it came on stream in 1980, Edmund Dell, then Labour's Trade Secretary, argued for a more protectionist policy against the Third World to try and ensure that thousands of others would not join the 80,000 jobs that had been lost in the British textile industry over the previous four years. More than £80 millions of public money was then spent on the de Lorean luxury car plant in Northern Ireland, a project relying on a high-priced product and an export market. Were there not scores of other products those skilled people would have preferred to make?

Employment

In 1851, industry for the first time employed more people in Britain than agriculture. Less than a hundred and fifty years earlier, 92 percent of the labour force worked on farms. By 1978 that figure had been cut down to less than 3 percent (and in the United States, about three percent of the labour force supplied nearly all the country's food needs as well as substantial exports of food). Professor Tom Stonier, head of the School of Science and Society at Bradford University, told the Government's Think-Tank that within thirty years Britain will need no more than 10 percent of its labour force to supply all its material needs.

The level of unemployment in Sunderland that year — 1978 — in the town that was once the largest shipbuilding centre in the world, was among the highest in the country. The city council was now the largest employer. Merthyr Tydfil, in South Wales, was once the home of the world's finest and largest ironworks. I invite you to visit it now. Go, too, to the valleys of South Wales where tens of thousands of steelworkers are being made redundant, and where the grandchildren and great-grandchildren of those who provided the sweated labour to make Britain the workshop of the world, are being thrown by the wayside as their ancestors were dragged from the country areas of mid-Wales years ago. They are treated as labour statistics, rather than human beings.

Over two million jobs in manufacturing industry disappeared between 1970 and 1980, a time when "the labour force" was increasing. One million of these went in two years, between 1979 and 1981: sixteen percent of all manufacturing jobs. There are now less than six million people working in industry. In the meantime, the service sector and white-collar office workers in local and central government enjoyed a boom. The service sector accounts for half of all jobs, compared to about 30 percent in 1964. During the ten years up to 1977, town halls and Whitehall employed an additional 1½ million people. Oxford City Council employs more people than the British Leyland plant up the road in Cowley. The National Health Service payroll has more than doubled over the past ten years and is now estimated to be the largest employer in Europe, with a total staff (in 1980) of over 900,000. Over that period, the number of hospital beds fell.

So while the number of "unproductive" people on the payroll and bureaucracies swell, services diminish. Resources are squandered: according to official estimates at the end of 1981, each person unemployed cost £4,380 in benefits and lost taxes and that did not include the benefits of increased production. Unemployment was then costing about £12,000 millions a year. Meanwhile, the revenue from North Sea oil taxes — £6,000 millions in 1981, rising probably to £10,000 millions in 1985 — is merely another addition to the sinking fund to reduce the Government's borrowing needs, which are increasing all the time, if only to pay for social security benefits for the unemployed.

The prevailing attitude of the Establishment is simply one of anxiety about the threat all this poses to the status quo. It is epitomised by the words of Sir Charles Carter, vice-chancellor of the University of Lancaster, chairman of the research and management committee of Britain's new think-tank, the Policy Studies Institute, and now a member of the Social Democratic Party. "Unemployment", he says, "is a key problem since it is a source of possible social unrest."

The notion is defeatist that since we can no longer sell what we used to, we might as well put on a clean shirt, if we can afford one and work in an office or accept social security handouts. It is misleading. As the authors of *Nature's Price*[1] put it:

the primary aim of production is not to keep up with foreign countries but to maintain a desirable level of prosperity". Moreover, the prevailing attitude assumes that everything society needs will be readily available, an attitude that is not only singularly materialistic, but one that avoids the whole question of who decides what should be produced and why. The most obvious example is arms production, and one of the most eloquent challenges came from workers on the shop floor: the Lucas Aerospace Combine Shop Stewards Committee in their Corporate Plan.

The Plan was drafted in 1976 following heavy redundancies, and the threat, since realised, of further lay-offs. It is a plan to provide alternative "socially useful" work. Automation and technological innovation, the shop stewards pointed out, had meant the sack rather than the promise of a shorter working week better working conditions and more leisure time. The Committee rejected the Labour leadership's support for steadily-increasing defence expenditure and its argument that the choice was between more and more military production or the dole. "Among the unemployed", they said, "are thousands of highly skilled engineers at a time when we urgently need cheap urban transport systems. Thousands of electricians are in the dole queue when we need cheap, effective heating systems. And there are thousands of workers in the dole queue when 7 million people live in semi-slums in this country and there is a tragic need for more hospitals and schools."

The Committee was formed after the "rationalisation" fashion that seduced British industrialists. For example, Arnold Weinstock, head of GEC, reduced his workforce from 260,000 to 200,000 at a time when that company's profits rose from £75 millions to £108 millions in one year. When the Committee launched its Plan, it said: "There is something seriously wrong with an economic, political and scientific framework which can produce Concorde, yet in that same society old-age pensioners are dying of hyperthermia for want of a simple, effective heating system".

Lucas Aerospace produce parts for military aircraft; it also makes kidney machines. But only enough to satisfy what the company calls "demand". 3,000 people in Britain die each year because they cannot get a kidney machine.

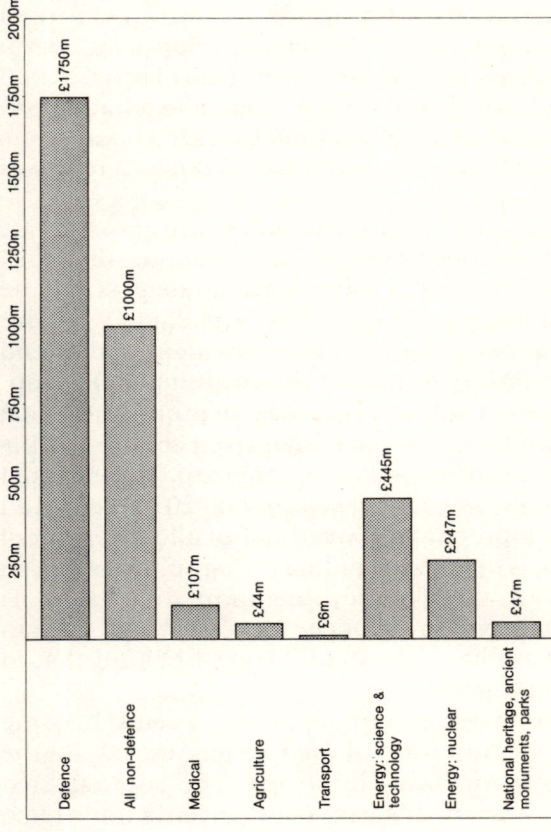

26. **Government Funds Allocated to Research:** *In 1980 the Government abolished the post of chief scientist at the Ministry of Agriculture – an indication of its priorities. All other important Whitehall departments, including the Ministry of Defence, has one. The figures do not include the money spent on "agricultural improvement", a euphemism for farm modernisation, which is about 50 times as much as it spends each year on conservation. (Source: Govt. Public Expenditure 1981-.)*

In 1978, the Combine Committee set up a Centre for Alternative Industrial and Technological Systems in cooperation with the North East London Polytechnic and the help of a small grant from the Joseph Rowntree Memorial Trust. It planned to help set up local, East End worker cooperatives. At least that is one challenge to conventional market mechanisms and prevailing attitudes.

One of its projects is a hybrid road/rail vehicle. This aroused the interest of the Highlands and Islands Development Board. It could also have provided Tanzania with a bill of just £20,000 per mile instead of the £2 million per track mile it had to pay for its new Uhuru railway: a railway built by China because the World Bank insisted that, as far as it was concerned, a road was the answer.

Military expenditure and arms sales are persistently defended by the Establishment on the grounds that they provide jobs. Yet there is plenty of evidence to suggest that the reverse is the truth and that countries with higher relative military spending have higher unemployment rates, and lower growth rates. Mary Kaldor of the Institute of Development Studies at Sussex University has demonstrated that Britain is the second most science-intensive economy of the West (after the US) as measured by the share of GNP devoted to expenditure on research and development. 20 percent of all research and development, and over half of all Government spending on R&D, is devoted to military programmes. Britain is the only country in the West devoting more than half of its public research budget on defence. Since the war, civilian innovation in Britain has consistently lagged behind that in other developed countries.

As capital costs increase and weapons systems become more complex, the number of people employed in the construction of weapons actually drops. The short-sighted approach by the Government's arms salesmen is demonstrated by the fact that just before he fell from power the Shah of Iran was by far the most important customer of the Ministry of Defence Royal Ordnance Factories.

It is widely acknowledged that by channelling so much investment and jobs to the military market in the shipbuilding and engineering industries during the early 1950s Britain lost an important place in the world market. Mary Kaldor argues: "Military spending can be viewed as a dynamic element in a process of decline, representing a Government response to decline — to unemployment or to threatened bankruptcy — which generates an internal momentum, which in turn absorbs resources that might otherwise have contributed to investment and economic growth". Military spending accounts for 20 percent of mechanical engineering output, half of ship-building output, and about threequarters of aerospace output. Military spending, Kaldor argues, can therefore be seen to postpone economic crisis and collapse, and also to prevent change. So unless alternative work for military industries is found, Britain is doomed to devote an increasing part of its GNP to military use, and encourage purchases of sophisticated military equipment in Third World countries with all that entails — diverting precious resources from initiatives of benefit to the whole of the community.

Even that part of the Establishment that is supposed to promote the interests of the mass of working people, the Trades Union Congress, has attempted to block attempts by

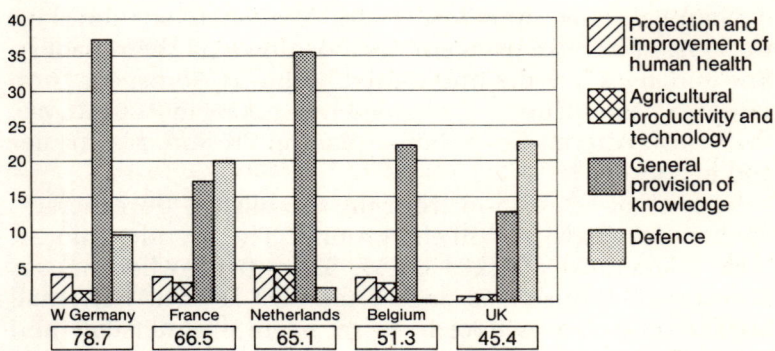

27. International Comparisons of Research: *The figures are for government expenditure in 1977 per head of population (in E.U.A.). The British Government is the only one in the West which devotes more than half of its total research and development budget to defence.*
(Source: Sir Hans Kornberg, Univ. Sheffield lec., 1980)

groups of factory workers to cooperate across traditional union boundaries in working out alternative, more rational and long-sighted investment and production plans.

We are back to the question of production, what we produce and for whom. There is a glut of steel, yet millions of people do not have sewing machines, bicycles, farm tools. But since we are critical of the present international trading system, why should not developing countries produce their own steel (which is not to say that the Third World should produce the same amount of steel, per capita, that is consumed by the US, something which will take them 500 years with their existing production facilities). Britain should gear its production primarily to its own needs. That is why it is absurd for British Leyland to continue producing a whole range of cars. It is far from original to suggest that it should have produced, instead, such necessary and simple things as hand pumps for the Third World, at least for as long as it so desperately needs them. Nearer home, it could have produced more Landrovers, or even electric cars.

The blue collar workforce is as frustrated as its critics about successive Governments pumping hundreds of millions of pounds into "a bottomless pit" as Merseyside workers put it. National trade union leaders and Whitehall look down on workers, rather like news editors look down on their readers. But initiatives like the one started by Lucas Aerospace shop stewards have sprung up, planned by workers in Rolls Royce, the British Aircraft Corporation plant in Preston, at Chrysler and at Vickers.

In spite of opposition from the established bureaucracy, cooperatives have got off the ground. The Meriden motorcycle cooperative, backed by Mr Tony Benn when he was Minister of Industry, is one of the best known. But small worker cooperatives have been set up in Skelmersdale and Milton Keynes with the help of the Job Creation Programme. Wandsworth, Lewisham, Newham and Fife councils have encouraged cooperatives. There were in 1981 about 5,000 worker cooperatives and more than 500 agricultural cooperatives. In agriculture, a system of regional cooperatives grouping mixed farms have the additional advantages of breaking the power of large agribusiness corporations and of promoting sound, labour-intensive husbandry.

Farm cooperatives are far more widespread on the Continent than in Britain. In France, farmers share machines and each other's labour during harvest time, and even share an interest in freezing and packing plants and retail shops. In Britain, too many farmers are either complacent or arrogant, doffing their cap to some spurious notion of individual independence. In fact, they become more and more dependent on large food processing companies for their sales and on agribusiness corporations for their input needs. As Einydd 'Tom' Thomas, managing director of Eastern Counties Farmers Ltd., has said: "Farmers are big spenders. But the cash spent by farmers is ending up in the pockets of a few big concerns in fertiliser and feeding stuffs, machinery and chemical industries."

A system of worker and farmer cooperatives could spread and redistribute wealth, promote both the freedom and satisfaction of the individual and the interests of the local community. There is no doubt that the success of the Mondragon cooperative in Northern Spain owes much to local Basque pride, something that many of the visitors from Britain who go there acknowledge.

The State

The blame for unemployment and the abuse of work must be placed on the shoulders of Whitehall, the TUC and British management — the discredited troika of the corporate state — for failing in their important, and self-appointed task of managing the country's economy. We should, therefore, question both the people who take decisions and the way they do so. Two crucial developments have taken place, almost unnoticed: Decision-making, and even influence on decision-makers, has been taken out of the hands of ordinary people and the democratic, parliamentary process. An insular British establishment has failed to learn from what is going on in other countries, or consider shifts in the world balance of power and the pressure for change.

They do not acknowledge it — they cannot afford to — but leading politicians, backed up by an uncertain though powerful bureaucracy, are on the defensive. Despite the rhetoric, the Establishment wants "to play safe", leaving as many options open as possible. Thus, for instance, the Government accepts with alacrity the nuclear power programmes put forward by the CEGB, an expansionist and capitalistic agricultural policy, a policy that is designed to encourage foreign investment rather than promote its own home-based industry, and a defence policy that stretches both the economy and every branch of the Armed Forces to the satisfaction of nobody.

This uncertainty is both the cause and the symptom of the failure of those in Whitehall to think in the slightly longer-term, to draw up a list of priorities for Britain, its communities and its economy that can then be debated throughout the country. This uncertainty also encourages ministers to be seduced by the comforting womb of Whitehall and other international organisations, including the EEC, NATO's nuclear planning group, and the UN Security Council.

One of the most neglected of all subjects is the decline of parliamentary democracy and the myths perpetuated by textbooks, civil servants and teachers on how the constitution works. The story has it that civil servants perform merely the tasks desired by their ministers, their political masters. Furthermore, ministers, the fable continues, are responsible to Parliament for their actions and all the activities of civil servants. These myths have been shattered by belated admissions that 1. civil servants have tremendous discretionary powers — over immigration policy, for instance, or planning applications, and that 2. ministers simply cannot keep up with everything that is going on in their name. Partly because of the activities of powerful pressure groups. Partly because of the parliamentary whipping system, and because of lack of time and information, MPs cannot effectively control the Executive. Those few MPs who are not embraced and seduced by the promise of a place in government find it virtually impossible to find out what is going on and how thousands of millions of pounds are being spent in Parliament's name by the taxpayer.

Decisions are in practice taken by powerful groups of vested interests in cooperation with the bureaucracy. This is not unusual in industrial societies either in the Western or in the Communist world (state capitalism and a privileged bureaucracy are, of course, the hallmarks of the Soviet Union's system of government). This has led to a corporate state and though Mrs. Thatcher, with her particular brand of populist radicalism, has tried to break the mould, it is ironic that if the Social Democratic-Liberal Alliance does secure power it will seek to restore a corporate consensus. More and more decisions will be taken by "apolitical" technocrats. Civil servants are not subject to effective democratic scrutiny and public accountability; but they believe, unlike politicians, who by definition represent the views of only one party, and therefore only one section of opinion in the country, that they represent a "common ground" and "the wider national interest", phrases which come easily to them.

It is far away from the criteria suggested by Professor A. H. Halsey, one of Britain's leading social scientists, in his 1978 Reith lectures:

"Democratic politics is, essentially, a system in which citizens actively mould the final decisions binding on all. It works only if liberty of thought and expression is ranked first among rights, and the active exercise of citizenship first among duties. Political action is inevitably carried on by imperfect people in public office. Hence the constant need for an alert and knowledgeable citizenry to protect itself against oppression and to prevent the public services from disintegrating into organisations which serve the private interests of public servants. Bold advances towards wider citizenship — in Scotland for the Scots, in the workplace for workers, in the school for parents, in the locality for the neighbour, and so on, could evoke popular support. Of course, there is also opposition. There are always those who place private interest above public welfare. Patient persuasion has to be our principal weapon: for persuasion is democracy in action. Only if it is given constant priority can we both guard our freedoms and override the resistance of vested interest. Only then, but certainly then, has democracy the right to enforce its will."

The potential consequences of the abuse of parliamentary democracy, coupled with high levels of unemployment, are frightening. It is likely to lead to greater intolerance and authoritarian pressures. If you take the view that any move towards the re-establishment of a consensus, through a national government, for example, would have only a temporary, superficial, success, and believe that Britain is in the process of a long-term decline that began well over 100 years ago, then the prospects are bleak. Polarisation between "right" and "left" seems inevitable. But there will be a less obvious, but a frightening and even more important polarisation between authoritarianism and libertarianism.

This could lead to what some have referred to "public sector imperialism". The strength of the public, or state-employed sector in the trade union movement is growing; more than half the members of the Trades Union Congress are in the public sector. Partly to counter this there is likely to be more pressure, not only against great bureaucratic, centralised, organisations representing particular groups, but for *less government* as well. At a time when those who argue for a stronger State — especially to face up to "law and order" — are becoming more and more shrill, the divisions between the authoritarians and libertarians will widen. The public sector is expanding; it is now managed by a breed of politbureaucrats, but it is not productive either. The State spends half our money; it expands its territory, but services to the public do not improve. Policemen are paid higher and higher wages relative to the rest of the community, and their numbers increase, yet crime rates also increase.

In policing, as in housing, education or policy towards the environment, the Government and established pressure groups are attacking the symptom, not the cause. Experience — surprisingly called an "experiment" even though it is based on ancient custom and tradition — in Devon and Cornwall demonstrates that where the community is at least understood, where police identify with and consult people to whom they are ultimately responsible, they actually discourage crime. Less law, more order. Very little attention is paid to the causes of discontented youth and run-down city areas, to why children of willing but deprived parents (poor and uneducated and so fewer books in the home) do less well in school despite "equality of opportunity", to why there is a severe shortage of nursery schools, how to combat desperate environmental and living conditions. Alienation obviously leads to an anti-community attitude, a kind of self-centred individualism, the black economy, an attack on all forms of "bureaucracy" and public services, without any constructive attempt to discuss what should be the role and status of the public sector.

If the State does nothing for us, why should we respect it and all its works? Is the State the sum of its citizens; what is the State without us? The questions are rarely posed. What resources does the State have if not ours? It has to rely on borrowing from the City which, like any lender, demands interest. The elected Government thus has no say over the level of interest rates and little or no say over the exchange rate of the pound. Meanwhile, the pool of money belonging to the State, that is to say, to us as citizens, consumers and taxpayers, is distributed arbitrarily, wasted, or earmarked to those members of the most influential pressure groups, regardless of the wider need and with very little rational discussion about priorities or discussion about values.

And the conventional wisdom, still, is that economic growth is sustainable. It is, after all, easier to promise a better standard of living for all, including the poor, if the pool of money is getting bigger. The easiest and simplest counter-argument is that however much the global increase in economic wealth under the present system, it will never be distributed equally, fairly, or justly. Indeed the gap between the privileged and the underprivileged is widening.

The dilemma was well put by Fred Hirsch in *The Social Limits to Growth*. He did not attempt to deny that economic growth had brought material gains. His point was that it failed to bring benefits, or contentment, in equal measure. Ambitions generated by the pursuit of growth are directed precisely at those things that are not available to everyone because their attraction is that they are relatively scarce: a country cottage, for instance, or an old master (or, now, a video). David Gordon said in *The Economist* after the death of Hirsch in January 1978 (at the age of 46) that Hirsch was about to enlarge his thesis and consider inflation. Could it be demonstrated that inflation was partly due to the failure of growth to meet the desire for greater distribution that growth itself provoked? In that case, if there is growth, you change the system. If there is no room for it, you devise a consumption pattern that is not inflationary. For what Hirsch was saying in so many words is that the more there is, the greater the expectation. In our present system, that greater wealth — in the broadest meaning of the term — is lost. Or, at least, if it finds its way into useful channels, even into the development of new technology, it is by luck, rather than by design.

Halsey, who is now very much out of favour with the Establishment, said in his Reith lectures that in a political democracy, the paramount principle of distribution must be equality. Equality of opportunity is not enough. We need full equality of material conditions, equality of income in the broad sense, as the foundation of social life; full equality that is modified only by such extra rewards to effort and capacity as can be shown to be necessary for "an efficient division of labour". I would add: "For those communities or individuals who needed those rewards to contribute to the general good".

And since we are not turning our back on progress, let us take new technology. Who is going to enjoy the benefits of mini-computers? Just those who have not only the necessary financial resources, but also the knowledge? Will there be no connection other than names on computer tapes between the bureaucrats, technocrats and police and the rest of us living, perhaps, our own lives in small groups in which personal relationships, voluntary or part-term work will play a much more important role?

An optimistic answer is given by Ivan Illich. In 1968, he says in *The Right to Useful Employment*: "It was still quite easy to dismiss organised lay resistance to professional dominance as nothing more than a throwback to romantic obscurantist or elitist fantasies. The grass roots, commonsense assessments of technological systems which I then outlined," he says, "seemed childish or retrograde to the political leaders...and to the 'radical' professionals who laid claim to the tutorship of the poor by means of their special knowledge".

Now, he says, the picture has changed. "A hallmark of advanced and enlightened technical competence is a self-confident community, neighbourhood, or group of citizens engaged in the systematic analysis and consequent ridicule of the 'needs,' 'problems', and 'solutions' defined for them by the agents of professional establishments." People have discovered, he suggests — and I wish he was right — what vital statistics have always shown: adult life expectancy has not changed in any significant way over the last few generations, is lower in most rich countries today than in our grandparents' time, and lower than in many poor countries. He observes "time-consuming acceleration, sick-making health care, stupefying education".

Males in particular *are* beginning to die at a younger age, on average, in the richest nations[2]. In Britain, life expectancy for men in the period, 1968–70, was 68.5 years compared to 68.7 in 1963–65. The US Senate Committee on Nutrition and Human Needs was told that "the life expectancy of the American male is actually decreasing". Diet and pollution, including smoking, are the main causes. The British Association for the Advancement of Science, referring to figures that suggest food poisoning may be increasing, mentions the use of intensive methods of livestock farming (designed to produce more meat for human consumption while reducing labour costs) as one possible cause.

Let us quote here the case of the Japanese lieutenant, Hiroo Onoda, who survived for thirty years in mountainous Philippine jungle before finally surrendering in 1974. He was reported to be "one of the healthiest 52 year-old men in Japan". During his stay in hospital he astonished the doctors who performed some 200 tests on him. Despite his ordeal, he had few defects and, indeed, was in far better physical and mental shape than most Japanese living in modern, urban affluence, suffering from pollution and general nervous strain.

Charging around in a polluted, hectic, environment is not doing our health any good. The poorer you are, the more likely you are to be vulnerable to disease. A clear link has been established between unemployment and poor health. According to conservative estimates, diet-related diseases cost the National Health Service at least £850 millions a year. Constipation alone cost the NHS £20 millions in 1979[3].

But Nature, too, has a price. Since we have always regarded Nature as free, we have not attached any monetary value to it. Environmental "products" — lakes, woods, water, fresh air — are not available on the market. Or are they? Access to them certainly is. What is more, we can put a clear value on polluting and damaging Nature. There is a cost attached to polluting lakes, for instance, to building expensive substitutes such as swimming pools or recreational activities because there are not enough natural habitats for us to pursue our pleasures.

Guided by that universal, conventional, measure of economic success, the Gross National Product, no respecter, more an exploiter, of Nature, we are plundering the earth. We are eating into natural capital.

"At the beginning of life there is a process called carbonic acid assimilation or photosynthesis."[4] Solar energy is thereby converted into a form which can be used by living creatures. By poisoning the earth and the atmosphere, we are threatening to kill that process at birth. Part of the trouble is simply that we still take Nature and the land for granted. Indeed, we are positively encouraged to do so. But, in the end, it is the only thing we have got.

And we still have not got used to looking at our own island. For hundreds and thousands of years, people worked on the land, in the countryside. Now more and more people are visiting it, escaping to it from their urban environment. They will have to make up their mind whether they want to keep it, or allow it to be spoiled so that they have nowhere to go.

We need space. Alice Coleman described the search earlier this century[5]: "The first-comers in the countryside near towns saw their peaceful spaciousness become congested with like-minded commuters and their vistas of rural beauty degenerate into weedy wasteland which often attracted rubbish dumping. The second wave of dreamers rejected this dreary prospect and moved on into the next zone of unspoilt farmscape, there to initiate the same wasteful and destructive procedure."

But there is still enough room so long as we respect the land and the countryside, do not allow road lobbies, greedy landowners, and others to ruin it.

*   *   *   *

POSTSCRIPT: ELEMENTS FOR A PROGRAMME

Democracy, politics, and institutions

Established politicians and senior civil servants, foxed about what to propose against the background of a Britain in economic decline, are on the defensive. Because they are worried and have little faith in themselves, they are concerned more about presenting themselves as people confident in what they are doing and give more priority to rhetoric than to serious debate about the substance of policy. They have been forced into cynicism, the refuge of those who are uncertain yet ambitious.

In turn, popular cynicism about professional politicians has increased and has been encouraged by the myths surrounding our hallowed institutions; above all, the House of Commons whose intricate procedures, more appropriate for a bygone age, receive more attention than the substantial issue of how to make Government more accountable.

But there are straws in the wind: pressure, mainly through minority parties like the Liberals, or new parties, like the Social Democratic Party, for a system of proportional representation (PR). It is a movement that would weaken the stranglehold of the traditional party whips and of the strict discipline of the mass parties in the Commons.

At the same time, backbench MPs of all parties are increasingly frustrated about their lack of control — and financial control above all — over the Executive: the combined forces of ministers and the mandarins of Whitehall.

But PR could well lead to a weak Parliament with no party having a majority. That may lead, in turn, to government by a centrist consensus with decisions in practice taken by a

corporate State: the Confederation of British Industry, the
Trades Union Congress, large companies and influential
pressure groups which are accountable to no-one, with the
possible exception of their shareholders. So new parties, or
PR, will not by themselves do any good.

Political institutions must be made to restore their links with
the people and local communities. Political party programmes
and manifestos must be drawn up and debated during an
election campaign. But they must also be carried out if the
party, or parties, which earn convincing and widespread
support are to be respected.

We do not need muddled government, or even less government, just more accountable government and more debate.
The concept of The State must be taken off its pedestal.

A second chamber, a reformed "House of Peers", half of its
members elected by regional assemblies, half co-opted by
their peers in different walks of life such as the arts or trade
unions and commerce and industry, would have a delaying
power over important legislative initiatives.

A Supreme Court, akin to the present Law Lords, would
consist of members of this second chamber.

Regional assemblies — for Scotland, Wales, Yorkshire, East
Anglia and so on, would be elected; half directly and half from
the members of local, district, councils. The assemblies would
be responsible for health services, education, large planning
programmes and policing. But minimum standards would be
set by the House of Commons which would also vote an
equalisation fund so that the poorer regions would not be
penalised.

Local councillors would be elected directly, with a third of
them up for re-election every year. They would automatically
receive a significant share of local rates although, again, there
would be a national equalisation fund. There would be little
use perpetuating and even encouraging economic disparities
between different areas.

Under these local councils, there would be parish councils,
with members directly elected, and which would have to be
consulted on all issues affecting their area and would have a
veto over proposals and issues that directly affected their
immediate environment.

Public inquiries into major planning schemes would be funded by the applicants, and both proposers and objectors would have the right to see all documents and technical data, free of charge, on which both sides based their case.

Financial institutions — mainly, the banks, insurance companies and pension funds — would be directly accountable to the House of Commons, which would scrutinise their investment programmes (though, in the case of pension funds, this role would be taken by existing and potential pensioners). Investment overseas of money earned in Britain would be strictly controlled by the Commons.

Main economic indicators and such benchmarks as the GNP would be either abandoned or redefined so as to reflect a fundamental change in values. They would reflect the needs of the community as a whole rather than free market forces and international capitalist interests. The fair distribution of wealth would be the prime consideration.

Agriculture

The automatic right to inherit estates would be abolished. Large estates with parks and gardens and efficient agricultural enterprises would not be broken up. Instead, they would be handed over to trusts or cooperatives managed by local communities. In those few areas where it makes clear economic sense to maintain specialised farming — in parts of East Anglia, for instance — the present farming pattern (though not necessarily the system) should remain. In other areas, mixed farming must be the rule, not the exception. There should be a system of cooperatives, with noone occupying more than any other. In any case, no individual should own or occupy land above a certain, agreed, limit (varying with the quality of land).

Though Britain could remain a member of a radically-changed EEC, it would make it clear that its own, national, agricultural, and indeed, other policies would be determined by criteria relating to national self-reliance rather than crude price and structural policies which bear no relation to the balance and needs of the British economy.

Housing

The right to a decent home, and a healthy and attractive environment, should be regarded as an inalienable right. With the exception of those for the elderly or infirm, all council houses should be sold off cheaply on terms which take into account the fact that the interest burden councils have to face on their housing account, the cost of housing administration and maintenance charges in most cases far exceeds the rent income from tenants[1]. Those families arbitrarily placed in tower blocks would be given houses now standing empty. (There is, after all, a surplus of about 1 million dwellings over households). House-swapping arrangements should replace the present open market for second homes. Tenancy for private dwellings would be abolished. Empty castles, of which there are several, should be inhabited by people.

Industry

Large enterprises should remain where it makes sense for them to do so. But they would be managed jointly by the workforce and the management grades, with the workforce having the final say over investment plans. Company and inter-company committees of shop stewards, regularly elected, would ensure that the national union bureaucracy — the TUC — would be reduced to a simple coordinating secretariat. At the local level, self-sufficiency would be the main criterion and the notion of an efficient, rational, international division of labour quietly buried. Worker cooperatives and job ownership schemes would be the norm, not the exception.

Education

There should be real equality of opportunity, as opposed to the paper principle — that every child, from whatever background, has the same opportunity. There would be one school system, with "public" schools becoming increasingly irrelevant. After the age of 15 or so, there would be sets within schools reflecting relative aptitude and subject interest. The importance of the social environment and parents' involvement in their childrens' education would be fully recognised, though appropriate national minimum standards would be fixed so that differences in various parts of the country — the pattern now — would disappear. The importance of adult education would be recognised, particularly in light of dramatic changes in employment conditions and life-styles.

Transport

High priority should be given to public transport, with railways subsidised, and road traffic penalised. Private cars in large towns and cities would be the exception, not the rule. Lorries would be forbidden to enter residential and urban areas unless they were on vital business, and delivering supplies. In a decentralised Britain, there would be fewer commuters and more local markets.

Law and Order

The police would be accountable to local councils and regional assemblies. Computers holding information on individuals would be open to the people concerned so they could check that records — needed for genuine medical reasons, for instance — were accurate. Information should be kept secret only in cases of proven need to fight crime.

Think-tank

Many more resources should be earmarked to the consideration of the country's future needs, planning priorities, economy, natural resources, and the implications of new technology. People from every background would be encouraged to participate in the process. Learning and discussion would help Britain to emerge from its present, singularly inarticulate, state.

*   *   *   *

NOTES

Introduction

1. Alice Coleman, "The use and misuse of our national land resources," *The Architects' Journal,* 19, January 1977, and Land Decade Educational Council, *Decade of Decision, Save or Squander?* 1980.
2. Angus Calder, *The People's War,* Panther Books, 1971.

Chapter 1. Ownership of the Land

1. E. P. Thompson, *The Making of the English Working Class,* Pelican 1975.
2. Peter Wormell, *Anatomy of Agriculture,* Harrap and Kluwer, 1978.
3. Royal Commission on the Distribution of Income and Wealth, report No. 7, 1979. (In 1981 the Conservative Government announced that it intended to stop all work on the collection of statistics relating to the distribution of income and wealth).
4. A. B. Atkinson and A. J. Harrison, *Distribution of Personal Wealth in Britain,* Cambridge University Press, 1978.
5. Ralph Whitlock, *Royal Farmers,* Michael Joseph, 1980.
6. see, for example, A. L. Poole, "Domesday Book to Magna Carta," in Oxford *History of England.*
7. Wormell, *ibid,* and A. Harrison, R. B. Tranter and R. S. Gibbs, *Landownership by Public and Semi-Public Institutions in the UK.* CAS Paper 3, 1977.
8. "Sheep on the Blasted Heath," *Farmers' Weekly,* 12 September, 1980.

Chapter 3. Marginal and Forgotten Areas

1. John McEwen, *Who Owns Scotland?*, Edinburgh University Student Publication Board/Polygon Books, 1981.
2. *The Observer*, 3 January 1982.
3. D. K. Britton and B. Hill, *Size and Efficiency in Farming*, Saxon House, 1975.
4. R. B. Tranter ed., *Smallfarming and the Nation*, CAS Paper 9, 1981.

Chapter 4. Concentration.

1. *Capital for Agriculture*, CAS Report 3, 1978.
2. *ibid.*
3. D. K. Britten and B. Hill, *Size and Efficiency in Farming*, Saxon House, 1975.
4. R. B. Tranter, ed. *Smallfarming and the Nation*, CAS Paper 9, 1981.

Chapter 5. Policy for Health

1. Calder, *ibid.*
2. see Elizabeth David, *English Bread and Yeast Cookery*, Allen Lane, Penguin Books, 1978.
3. *National Food Policy in the UK*, CAS Report 5, 1979.
4. *ibid.*

Chapter 6. Pests, Pollution and Waste.

1. Royal Commission on Environmental Pollution, seventh report, *Agriculture and Pollution*, HMSO, 1979.
2. see, for example, Barry Commoner, *Science and Survival*, New York, Viking Press, 1969.
3. see *New Scientist*, October 1981.
4. see *New Scientist*, March 1981.
5. *Farming Organically*, the Soil Association, 1979.
6. see Quentin Seddon, "The Mystery of the Decay," *The Guardian*, November 13, 1980.
7. Watkin Williams (professor), *UK Food Production: Resources and Alternatives*, *New Scientist*, December 1977.
8. Colin Tudge, *Future Cook*, Mitchell Beazley, 1980.

Chapter 7. The Planning Mess.

1. *Land for Agriculture,* CAS Report 1, 1976.

Chapter 8. The Mess on the Fringes

1. *Planning — Friend or Foe?* Council for the Protection of Rural England, 1981.

Chapter 9. Conflict

1. Geoffrey Wheatcroft, *The Spectator,* 12 May 1979.
2. Marion Shoard, *The Theft of the Countryside,* Maurice Temple Smith, 1980.
3. Shoard *ibid.*
4. M. Parry, A. Bruce,-and C. Harkness, *The Plight of British Moorlands, New Scientist,* May 1981.
5. *ibid.*
6. ECOS, the Journal of the British Association of Nature Conservationists, Vol. 2. No. 2. Spring 1981.
7. *ibid.*
8. F. G. Sturrock and J. Cathie, *Farm Modernisation and the Countryside,* University of Cambridge Department of Land Economy, Occasional Paper No. 12, 1980.

Chapter 10. Rural Deprivation and Country Attitudes

1. personal communication to the author.
2. *Sunday Times* magazine, 5 November 1978.
3. Child Poverty Action Group, *Rural Poverty,* November 1978.
4. Low Pay Unit report No. 3, November 1980, *Poor Harvest: Farmworkers and the Common Market.*
5. see notably the 1978 report by the Standing Conference of Rural County Councils.
6. Labour Party pamphlet, *Out of Town, Out of Mind,* 1981.
7. *The Spectator,* 21 March 1981.
8. H. Newby, C. Bell, D. Rose and P. Saunders, *Property, Paternalism and Power,* Hutchinson University Library, 1978.
9. Newby et al, *ibid.*
10. *ibid,* see also Howard Newby, *The Deferential Worker,* Penguin Education, 1977.

11. Newby et al, *ibid.*
12. Newby et al, *ibid.*
13. Agricultural Extension and Rural Development Centre, Reading University, *A Study of School-leavers entering Agriculture, 1979.*

Chapter 11. Third World Perspectives

1. *New Internationalist,* December 1976, article by Steve Jones.
2. Dan Morgan, *Merchants of Grain,* Weidenfeld and Nicolson, 1979.
3. Pat Mooney, *Seeds of the Earth, a Private or Public Resource?*, Inter Fares, Ottawa, 1980.
4. Mooney *ibid.*
5. Research by Harald Sioti, director of Max Planck Institute of Limnology.
6. The Economist Intelligence Unit, *International Trade in Grain and the World Food Economy,* September 1980.

Chapter 12. Now or Never

1. W. van Dieren and M. G. W. Hummelinck, *Nature's Price,* Marion Boyars, 1979.
2. see Malcolm Caldwell, *The Wealth of Some Nations,* Zed Press, 1977.
3. William Laing of the Office of Health Economics, British Nutrition Foundation annual conference 1980.
4. *Nature's Price, ibid.*

FURTHER READING

Bibliography

The Centre for Agricultural Strategy, based at Reading University, has provided a much-needed forum to question the prevailing attitudes in both the Ministry of Agriculture and the National Farmers' Union. The keen interest in its publications shown by the Ministry — on at least one occasion it intervened directly in an attempt to try and influence a report — is evidence of the Centre's success. Originally sponsored by the Nuffield Foundation in 1975, it was by 1981 suffering from a shortage of funds.

Some of its publications: Land for Agriculture, 1976. Capital for Agriculture, 1978. The Future of Upland Britain, 1978. The Economy of Upland Britain, 1750–1950 (1978). National Food Policy in the UK, 1979. Strategy for the UK Forest Industry, 1980. The Efficiency of British Agriculture, 1980. Self-sufficiency and Food Security, 1980. Smallfarming and the Nation, 1981. Grassland in the British Economy, 1981. Available from CAS, Reading University, 2 Earley Gate, Reading.

Of all the regular publications, I have found the *New Scientist* a consistent source of provocative, informative and enlightening material. It is especially useful for those, like the author, with no formal scientific education behind them.

Background to agriculture and rural society:

Anatomy of Agriculture, by Peter Wormell, Harrap and Kluwer, 1978.
The Theft of the Countryside, by Marion Shoard, Maurice Temple Smith, 1980.
Land For The People, compiled by Herbert Girardet, Crescent Books, 1976.

Property, Paternalism and Power, by H. Newby, C. Bell, D. Rose, and P. Saunders, Hutchinson University Library 1978.

On The Third World:

Food First, The Myth of Scarcity, by Frances Moor Lappe and Joseph Collins, Souvenir Press, 1980.
Seeds of the Earth, by Pat Mooney, available from Third World Publications, 151 Stratford Road, Birmingham 11.
The Famine Business, by Colin Tudge, Faber 1977.
How the Other Half Dies, The real reasons for world hunger, by Susan George, Penguin Books, 1976.

And more generally:

A Blueprint for Survival, The Ecologist, Penguin Books, 1972.
Progress for a Small Planet, Barbara Ward, Pelican Books, 1979.
Journey Through Britain, John Hillaby, Paladin, 1973.
Falling Apart, The Rise and Decline of Urban Civilisation, Elaine Morgan, Souvenir Press, 1976.
Nature's Price, The Economics of Mother Earth, Marion Boyars, 1979.
245-T, Portrait of a Prison, by Judith Cooke, and Christopher Kaufman, Pluto Press, 1982.

INDEX